FINDING RESILIENCE

CHANGE AND UNCERTAINTY IN NATURE AND SOCIETY

BRIAN WALKER

Dedication

For Laura, Deena, Sean, Kate and Ross

FINDING RESILIENCE

CHANGE AND UNCERTAINTY IN NATURE AND SOCIETY

BRIAN WALKER

CSIRO

PUBLISHING

CABI

A catalogue record for this book is available from the National Library of Australia and from the British Library, London, UK.

Published exclusively in Australia and New Zealand by:
CSIRO Publishing
Locked Bag 10
Clayton South VIC 3169
Australia

Telephone: +61 3 9545 8400
Email: publishing.sales@csiro.au
Website: www.publish.csiro.au

Published exclusively throughout the world (excluding Australia and New Zealand) by CABI, with ISBN 9781789241594

CABI	CABI
Nosworthy Way	745 Atlantic Avenue
Wallingford	8th Floor
Oxfordshire OX10 8DE	Boston, MA 02111
UK	USA
Tel: +44 (0)1491 832111	Tel: +1 (617)682-9015
Fax: +44 (0)1491 833508	E-mail: cabi-nao@cabi.org
E-mail: info@cabi.org	
Website: www.cabi.org	

Front cover: (top) Manhattan skyline and people on the lawn in Central Park (photo: Roman Babakin/Shutterstock); (middle) bramble (*Rubus* sp.) (photo: Anest/Shutterstock); (bottom) aerial view of Okavango Delta river (photo: Ingrid Heres/Shutterstock)

Set in 11/13.5 Minion and Helvetica Neue
Edited by Peter Storer
Cover design by Andrew Weatherill
Typeset by Thomson Digital
Printed in Singapore by Markono Print Media Pte Ltd

Contents

Acknowledgements

Of the many people I should acknowledge for the life and career I have enjoyed, I want to single out just two. Bob Coupland gave me the chance to go to the University of Saskatchewan and (especially) gave me the freedom to follow my nose in terms of what interested me for my PhD. Somewhat later, CS (Buzz) Holling introduced me to the ideas of resilience during a sabbatical with him in Vancouver. Our subsequent friendship and collaboration strongly influenced the intellectual course of my life.

This book unfolds as a pathway to understanding what resilience is and how it is expressed. It reflects my own journey, mostly played out through the activities and findings of students and a broad range of colleagues with whom I have interacted. I acknowledge them all for their insights and wisdom. In writing the book, Steve Carpenter, Fred Ellery, Tony Ferrar, Pete Goodman, Eric Lambin, Raphael Mathevet, Mary Seely and Sergio Villamayor Tomás provided valuable input to the content of sections in which they were involved or had particular knowledge. I hope all those mentioned will forgive the licence I have taken here and there – interpreting thoughts and in describing incidents, all of which were real.

The Universities of Rhodesia (now Zimbabwe) and the Witwatersrand supported me in all my research efforts in southern Africa and, for over 30 years, Australia's Commonwealth Scientific and Industrial Research Organisation (CSIRO) has provided me with greatly appreciated opportunities to explore new ideas.

The ideas and concepts towards the end of the book are drawn from the members of the Resilience Alliance, an international gathering of scientists across disciplines, and from the Royal Swedish Academy of Science's Beijer Institute of Ecological Economics: two extraordinary groups of people with whom I've had the privilege to be associated for nigh on 30 years.

As I tried to pull together the evolution of my ideas through the stories around them, it developed into a mix that was neither science nor interesting, and my chats over coffee with Jennifer Barton about writing and what makes for a good book helped me enormously.

Finally and most importantly, a huge thanks goes to my family from whom I have been away too much pursuing what's recounted in the book. They were always, all of them, in my mind as I thought about the future. To Deena, Sean, Kate and Ross for their comments on various excerpts, and especially, and with my love, to Laura, my wife, greatest supporter and most honest critic. My parents, who gave me opportunities they never had, made everything possible.

PART I
What's it all about?

What's it all about, Alfie?
Is it just for the moment we live?
What's it all about when you sort it out, Alfie?
Are we meant to take more than we give?[1]

1

Connections in a changing world

Where are we heading?

It's hard to grasp the enormity and timescales of the changes that have happened since our world began, and it's particularly hard to understand how they relate to what is happening now. One way of trying to do it has been by compressing the history of Earth into a year, highlighting the major events and changes as the year unfolds. An interesting version of this is to imagine a year-long movie of Earth, beginning at the time it first condensed into a globe. The movie would be made up from one composite picture per year portraying the whole surface of Earth in one frame. It would have around 4.6 billion frames (the age of Earth) and, if the speed of the projector is such that the movie begins at 1 second into the new year and finishes on the stroke of midnight on 31 December, the following is what you would see.

For the first 3 days in January, the Earth is a boiling mass in which denser substances sink into a molten core (which still exists). As the atmosphere thins and cools, the surface crust forms. Water vapour accumulates in the atmosphere, stabilising temperatures. Condensation leads to rain, the hydrological cycle develops and on 4 January the oceans form (all this took close to 50 million years).

At the end of January, *the* big event in Earth's history occurs as the first signs of life appear: tiny, simple organisms, primitive bacteria. Invisible at first, they multiply and diversify. The seas become full of these primitive bacteria, which develop and grow using energy from chemical interactions. They begin to change and become more complex as evolution gets underway.

In the latter half of May, there is another stupendous change. It begins with an imperceptible green dot inside a bacterium in the vast expanse of the primordial ocean soup: the first appearance of chlorophyll. This amazing molecule reflects green light and absorbs blue and red, using this energy to combine carbon dioxide and water into sugar, and releasing oxygen as a by-product. Evolution leads to multicellular organisms and in mid-June a new kind of chlorophyll first appears in green algae, and then in higher plant species as they evolve and diversify and invade the land. By early August, terrestrial Earth has become green, changing from having an atmosphere practically devoid of oxygen to one containing around 20%.

Over the next three and a half months (through to the middle of November), a great deal of biological activity is apparent with the evolution of millions of different species, and from 13 October until 13 November (400 million years), the world looks completely white as it goes through a long ice age. This is 'snowball Earth', with the coldest temperatures ever experienced. A host of species go extinct, but evolution takes off again and life diversifies. On 25 November, a flash of light is followed by a momentary darkness over the world as a murky layer fills the atmosphere. An asteroid has crashed into Earth and a host of species go extinct.

On 1 December, there is another asteroid impact. About 70% of species go extinct and late on that day the first vertebrate animals appear on land. On 11 December, the next asteroid impact occurs (the biggest catastrophe for life that Earth has known) causing around 95% of species to go extinct. On 12 December, reptiles appear and soon the age of the dinosaurs begins during the warmest period for life on Earth. Another catastrophic loss of species happens on 15 December, and on 18 December India collides with Asia, throwing up the Himalayas, and the first mammals appear, followed shortly by the first recognisable birds, as the dinosaurs continue to thrive in the middle of their 135 million year reign.

Late on Christmas night, the fifth (most recent) extinction event occurs and about three-quarters of the world's species disappear. The dinosaurs die out and an explosion of mammal species begins.

On 31 December, at ~4.30 pm, the first bipeds emerge in the subtropical regions of Africa: ape-like creatures walking upright on two legs. Just after 8 pm, the first member of our genus, *Homo erectus*, appears, and at 11.35 pm, just 25 minutes before the end of the year, *Homo sapiens* appears. This species spreads rapidly through the continent and some in the northern part of Africa soon cross into Europe, replacing (probably exterminating) the pre-human *Australopithecus africanus* populations that arrived there earlier. *Homo sapiens* radiates through Europe, evolving into different races adapted to different environments, and by ~12 minutes to midnight people have reached Asia.

After ~6 pm, the world goes through many glaciation episodes (17 in the last 2 million years)[2] and it reaches its most recent maximum glaciated phase at 2 minutes to midnight. Within 50 seconds, with just one minute and 10 seconds of the year left, the ice sheets retreat and Earth is again in a warm phase.

At 1 minute to midnight, the world is a beautiful sight – the continents are covered by forests, savannas, sweeping grasslands and deserts, seas teem with marine life, each with their uniquely adapted forms of plants and animals. With 30 seconds to go, some little blotches appear: the first signs of human settlements. As the seconds tick by, they grow in size and become more numerous. With 3 seconds to midnight, the Holocene epoch begins: the last 12 000 years when Earth's climate has been more stable and benign for longer than ever before, allowing *Homo sapiens* to really flourish.

It seems like the camera is speeding up and, with just 2 seconds to go, the Industrial Age arrives. Forests disappear, vast swathes of cultivated land appear, and the number and size of cities, and their murky palls, explode. As the sizes of cities increase, so, too, on a microsecond timescale, do the intensities of fires crammed together like a fireworks display. All this in the last 2 seconds of the year. It took more than 10 000 human lifetimes for the world's population to reach 3 billion (in 1960). It has since reached 7.5 billion within one lifetime.

Earth is now in the 'Anthropocene': an epoch when human activities are significantly affecting ecosystems and climate. And in the last fraction of the last second the rate of change is still increasing, ever faster and more extensive than anything since the movie began – barring the asteroid impact that obliterated most of life on Christmas eve. The movie ends abruptly.

If the movie could run on into the future for just one more second, what would we see? *Quo vadimus* – where are we going? Or, as Latin scholars would point out, a more accurate translation is 'where are we rushing to?'

Connections with nature

As primates, we have been co-evolving with the ecosystems in which we live for over 10 million years. We split from our nearest relatives, chimpanzees, ~6 million years ago and the genus *Homo* has been around for ~3 million years. The genes that govern our physiology and behaviour evolved over all this time to adapt us, physically, behaviourally and emotionally, to the environment and the ecosystems we live in. We are tightly linked to the nature of these ecosystems and to the other species that share them, from microbes to mammals, and a key part of our success has been our ability to sense and respond to these other species and to the environment in which we evolved.

Our bodies have evolved to connect us physiologically to the physical environment, although we are largely unaware of it and respond to its signals subconsciously. We accept without question the marvellously intricate mechanisms that allow our pupils to dilate and contract in response to changing light intensities and our bodies to maintain an even temperature. Our sensory and emotional connections are also mostly subconscious responses dictating how we feel and behave, but sometimes our conscious parts kick in and thinking about the signal can either enhance or reduce the response, depending on how serious it actually is. We can also be made aware of environmental signals as memories, which can strike us quite forcibly. I still vividly remember one such incident.

Just before sunset, in the hazy light of an African evening in spring, I drove down the escarpment of the Zambezi valley into the Mana Pools game reserve. I was with a botanist friend, Tom Muller, on our way to set up a research project on what was happening to the vegetation on the Zambezi River alluvium, following the construction of the Kariba dam. There was concern that the altered flow and flood regimes were changing the ecology of the alluvial ecosystems.

Having left the paved road, we were bumping and lurching our way down along a dirt road, and near the bottom of the escarpment we entered a woodland of flat-topped acacias. A warm feeling of familiarity flooded through me. It was my first return visit to such an area after 4 years in Canada. The shapes of the trees were so familiar, contrasting sharply with the thinner, erect shapes of pines and firs in the forests of northern Saskatchewan. The light was fading when we came out on to the broad alluvial flood plain of the Zambezi. Our way became smoother as we entered the magnificent open woodlands of winter thorn trees. We drove slowly through this paradise for several kilometres, catching fleeting glances of the grey shapes of elephants in the gathering gloom.

It was almost dark as we descended into a small dry riverbed, seeing at close quarters the long grass of the riverine fringe in the Land Rover's headlights. A band of cooler air accompanied by an unmistakable smell came through the open windows. I felt a prickly sensation, a thrill of familiarity and an absolutely wonderful feeling of being home.

'Ah! *Phyllanthus*' remarked Tom. In the gloom I nodded in agreement, savouring the moment.

Phyllanthus reticulatus is a shrub that grows along the banks of rivers and drainage channels in the low altitude subtropical regions, the 'lowveld', of southern Africa. In the still of the evening its small flowers open and, when cool air drains into the river channels after a hot day, the aroma of this plant fills the air in uneven bands and patches. It is like the smell of freshly peeled potato skins (giving rise to its common name, the potato plant) and for me it is the smell of Africa: the smell of that part of the world that means so much to me. Smells like this touch something inside most people who were born in and grew up in some part of the world with characteristic odours, and they trigger a psychological response that engenders a strong emotional tie with that environment. The Australian author Michelle de Kretser, who was born in Sri Lanka and lived her first 16 years there, described at a literary festival how smell was her dominant sense when she returned to Sri Lanka after many years, and still is: 'the fecund rot of the tropics makes me feel completely at home'.[3]

There is a lovely example of how this sense led to a breakthrough in understanding animal behaviour. Arthur Hassler was a renowned and much-loved ecologist at the University of Wisconsin. He died aged 93 in 2001 and it was he who answered one of the great intriguing mysteries of nature: how salmon returning from the sea to breed are able to precisely identify the stream, and the very place in that stream, where they were 'born', after being at sea for several years. In the 1940s, Arthur identified what is known as 'olfactory imprinting'. With their finely honed sense of smell, the very subtle differences from place to place enabled returning salmon to swim hundreds of kilometres, reversing the imprinted chain of smells recorded in their brains on the way down to the sea, to arrive back at the precise stream where they were spawned. It was on a visit back to a mountain stream in Utah where he grew up that Arthur noticed how the smells of the plants around him rekindled childhood memories. And he wondered if that happened to fish. It was the germ of an idea that eventually led his research to unravel just how finely tuned olfactory imprinting has become in salmon.

Sight and smell are just two of the ways we sense and respond to our environment: all five of our senses are capable of eliciting strong responses that identify and tie us to the place we were reared. Collectively, our responses to the environmental stimuli detected by these five senses strongly influence our mental wellbeing. Ask someone to identify, for each of the five senses, the thing that characterises best their part of the world, and you will get an interesting insight into that person. While the smell for me is the aroma of *Phyllanthus,* for a friend who spent his life in savanna rangelands it was the warm, moist smell of soil after the first rains. Sound for me is probably the gentle call of an African scops owl at night or, nowadays, after many years listening to them, the warbling of the Australian magpie in the early morning – as children's author Pamela Allen so aptly captures it, the 'waddle giggle gargle paddle poodle' instantly recognised by all Aussie kids. Hippos grunting in the river and belly rumbles of elephants are great reminders for me, but, as evocative sounds, these bird calls pip them.

The positive reactions triggered in our five senses by environmental signals are those that identify us with the places we feel we belong, that we understand and where we are comfortable. However, we also experience strong negative reactions in each of the senses, and these are much more general. For good reason, wherever we were reared and now live, we all have a strong aversion to the smell and taste of rotting meat. With a few exceptions, we are scared by the looks and hissing sounds of snakes. The scream of someone in great pain thoroughly alarms us. The painful experience of touching a burning ember, or being hooked by a sharp thorn, makes us wary of them in the future. Our senses both connect us positively to our environment and protect us from it.

But our five senses are being dulled in civilised man. For our sense of smell we have 5–6 million receptor cells high in our nasal passages (dogs have over

200 million) and we can potentially distinguish more than 10 000 different odours. But we can't all do that. Not only do we differ genetically in how well we can potentially smell things, but, for many of us, our potential sense of smell is poorly developed. We have not probed our sensory abilities nor used our senses in wide ranging ways. The old adage 'use it or lose it' applies well to the use of our senses. The average city dweller of today contrasts sharply with the remarkable abilities of Kalahari Bushmen and Australian Aboriginals who have grown up in natural ecosystems: not just living in them but living as a part of them and reliant on them. A colleague, Doug Williamson, once described to me an incident that captures this contrast.

He was in an open Land Rover on a sandy track in Botswana on the way to Deception Pan in the Kalahari. Behind him was a young San (Bushman) girl about 10 years old, sitting next to her father. They were travelling in a four-wheel drive at ~25 km an hour and the girl was leaning on the side of the Land Rover looking down at the road. Suddenly she called out to her father in an excited voice, in their distinctive 'click'-sounding language. The driver stopped. They all got out and walked back to check something in the track. The girl's father confirmed that yes, indeed, it was the spoor of uncle so-and-so. Uncle so-and-so hadn't been around for some months, which was why the girl was excited. Neither of the white men in the vehicle could make out the distinguishing marks the Bushman pointed out. It was difficult enough to see that it was a footprint. On arrival at Deception Pan, however, there was uncle so-and-so, full of smiles and not in the least surprised his niece had identified his footprint while travelling at 25 km an hour.

Our emotional responses to environmental signals are no less intricate than are our physical and physiological responses. We share the emotional part of our brains with all other vertebrates – the oldest ancestral part, the stem at the base of our brain, sometimes called the reptilian part. The large conscious (human) part of our brain is for rational thought. Our essential 'humanness' depends on both conscious and subconscious responses to environmental stimuli. The two parts of our brains are strongly connected and the growing numbers of emotionally ill people in the developed world highlight how important it is to do something about the connection. The emotional responses triggered by environmental stimuli confirm that we have evolved as a part of the ecosystems in which we live, and in all the truly amazing complexity of the human brain our emotional wellbeing is still responsive to, and in fact dependent on, the 'ecosystem' we're in.

This does not mean we have to stop all progress and keep ecosystems in exactly the state in which humans first saw them. In fact trying to do that – the preservationist philosophy – is a recipe for disaster. There are many changes we can make without upsetting our emotional links to ecosystems. Few people, ecologists included, could tell if they were in a pristine ecosystem or not. Three hundred years ago, much of Scotland was forested and looked like most of Sweden does today. People transformed it by clearing the forest. Water levels rose as a consequence, the forests were replaced by heathlands, and the present Scots love and identify with the heather and heathlands that now predominate.

What hasn't changed is humankind's basic relationship with ecosystems. We are all part of and connected with the ecosystems we live in. They may have the essential characteristics of the kinds we evolved in and affect us in positive ways, or they may have been so radically degraded or changed as to affect us negatively. Living in a congested city can do that. Either way, we respond in an intuitive way to the environment we're in.

Our response to the stimulus of being attacked by a predator, for example, has evolved over millions of years. It's a highly complex mechanism involving a sudden spurt of adrenalin into the blood that gives an instantaneous increase in heart rate and puts our bodies into the best possible state for rapid flight or fighting back. The problem in modern society is that this response is triggered by all sorts of annoying situations, such as friction with neighbours, colleagues and road hogs, and under these conditions what we really need is enhanced frontal brain activity and clear thinking rather than a physical response. Instead, our mental capacities are subdued and reduced in favour of getting our legs and arms working. A shot of adrenalin reduces the ability to think.

When confronted with an imminent attack, an early hominid who stood around wondering 'Should I fight back? Perhaps not. I think on this occasion I might flee, but then on the other hand …' didn't get to pass on his genes. The 'attack' response nowadays, however, is more often than not an inappropriate reaction to the stimulus. It is not only unhelpful, it leads to a wide array of stress-related illnesses illustrating that we modern humans are still genetically much the same as our *Homo habilis* forebears – it is our physical and social environment that has changed, not us.

But in one way we humans really are special when it comes to our big brains. We're the only species capable of foresight and reasoning. One of the top researchers in this area, Daniel Kahneman, sums it up by saying we have two systems in our brain: a fast one and a slow one. The fast one is our ancient, intuitive one that works on cues we receive – such as the fight or flight response – and our slow one is our reasoning system. His entertaining book *Thinking, Fast and Slow*[4] has some great examples of how the two systems can get in each other's way, leading to all kinds of problems. Many of the things we do are triggered by our fast system without us even being aware of it, and mostly that's good, because it saves a lot of time and unnecessary angst. But there are times when we need to deliberately engage our slow system and work things out before acting.

Not all of our responses to modern world stimuli are negative or inappropriate. Positive human values are triggered by music and other human-made environments and so are important for our wellbeing. This is not surprising because it was human brains that composed the music and designed the gardens that make us feel good. Research has shown that 'good' smells, pleasing music and 'happy' thoughts can have positive effects on our immune systems, while the opposites of these induce a lowering of immune-related proteins in our blood.

As the only species capable of self-reflection, and therefore of determining our own destiny, we need to build on the connections that evoke the sorts of emotional responses we value and that are consistent with what biologists call an 'evolutionary stable strategy'. We need responses that are long lasting and likely to favour our safety and wellbeing rather than irrational anger that puts us at risk.

Our connections with nature are not only to the ecosystems we live in but also to the ecosystems within us. The human body is not just a set of organs that work together to make it function as an organism. It is itself an ecosystem comprised of multiple organisms. Each of our bodies is inhabited by trillions of other organisms encompassing thousands of species, and the relationships between them and our bodies determines our wellbeing. They live on and in every imaginable part of us. On the outside ectoparasites – ticks, fleas, lice, mites and the like dominate. On the inside there is a mixture. There may be many parasites (endoparasites), such as flatworms, roundworms, tapeworms and hookworms. They seldom kill their host but they lower performance, reducing quality of life. Bilharzia, a small blood fluke, attaches itself to the inner walls of blood vessels where it can live up to 30 years, and afflicts millions of people in Asia and Africa. Roundworms and hookworms cause anaemia and diarrhoea in billions of people. And some parasites are lethal, such as the malaria parasites that infect red blood cells and claim the life of a child somewhere in the world every 30 seconds.

The debilitating effects of parasites make us very aware of them, but we are largely unaware of the enormous beneficial effects of all the symbiotic organisms inside us. It may surprise you to learn that 9 out of every 10 cells in your body are bacteria, most of them living symbiotically. Just as cows, deer and all other ruminant animals could not process the food they eat without them, you would not be able to properly digest your food without the bacteria in your digestive tract. And if you lack the right mix and numbers in your colon, you'll have great difficulty trying to poo, or you'll have diarrhoea. An emerging treatment for people suffering in this way due to an infection with a nasty bacterium is a faecal transplant – taking poo from a healthy person and inserting it via an enema into the colon. Our intestinal bacteria are critical for nutrient metabolism, they help protect against pathogens, support our immune systems, protect against colorectal cancer and affect neurological behaviour. Many diseases have been linked to malfunction in our microbial metabolism: cardiovascular disease, obesity, diabetes, irritable bowel disease, asthma and others.

Of the microbes that are purely parasitic, some, such as the smallpox virus, evolved a severe strategy that killed their hosts. This was still effective from the virus's point of view because there were always enough humans left to keep both species going, but, from our point of view, the relationship was definitely too one-sided, which is why we eventually eliminated smallpox. We have yet to achieve that with the malaria parasite.

We have become so accustomed to the spectacular effects of antibiotics and other medicines that most people no longer regard parasitism and disease as a real threat. But the rise and spread of new diseases such as HIV, SARS and the bird and swine flu viruses, plus the rising incidence of dengue haemorrhagic fever and Lyme disease, and the outbreaks of Ebola virus in West Africa and Zika virus in Uganda, later spreading from its appearance in Brazil where it was first identified with microcephaly and Guillain-Barré syndrome, tells us that they most surely are still a threat. Coupled with the serious problem of increasing drug resistance and emergence of 'super' pests and diseases such as malaria and tuberculosis in the developing world, and the rise of golden staph in developed world hospitals, these parasitism and disease trends serve to remind us that not only are we connected to the ecosystems we share with other species, but that our own bodies are ecosystems. We're still learning about the intricate and highly dynamic ways in which the many different kinds of organisms interact and change in response to the food we eat and the environment we're in. We are largely ignorant of how changes in our external environment might cause changes in our microbiomes, but, as the world's environment changes, it is highly likely they will, and therefore so will we – with consequences we can't predict.

Our connections to the physical environment, to the ecosystems we live in, and to the ones inside us, determine our behavioural, emotional and physiological wellbeing – our essential humanness. Our identity. We can adapt to quite marked fluctuations and changes in them, but there are limits. Even if we can forestall the apocalyptic collapse presaged by the end of the movie, beyond these limits we will not be able to function as the people we now are. There are limits to humanity's resilience.

2

Another pathway

Around 4700 years ago, Gilgamesh, the fifth ruler of the city-kingdom of Uruk in southern Mesopotamia, had some big tree-felling plans. He needed timber for ambitious developments in his kingdom – the original go-for-growth policy – and he began felling the cedar forests of southern Mesopotamia. What is now pretty much a wasteland was in Gilgamesh's day a vast area of primeval forest, and historian John Perlin has chronicled what happened to it in his book *A Forest Journey: the Story of Wood And Civilization.*[5]

Uruk is the area in what historically is known as the Fertile Crescent in present-day Iraq where Western civilisation emerged. It also happens to be the place where, according to Genesis, God created the Garden of Eden, clearly intended to provide sustenance for all creatures, not just us. Stretching the metaphor a bit, greed is the green-eyed serpent tempting us to take too much – the apple, from the tree of knowledge! We don't seem to have learned very much.

By the third millennium BC, many city-states in Mesopotamia were clearing timber in a big way. By the time of the Third Dynasty at Ur in 2100 BC, they were importing cedars from what is now southern Turkey. With the upper catchments and banks of the Euphrates, Tigris and Karun rivers stripped of their trees, soil erosion was rife and the rivers became silted and clogged up the irrigation canals. Salt began to accumulate and Sumerian scribes recorded huge drops in wheat production after 2300 BC. After more than a thousand years of successful agriculture, salinisation led to the decline of Sumerian civilisation. The last Sumerian empire collapsed in 2000 BC and the centre of power moved north to Babylonia.

The same sad saga unfolded around 1500 BC in Mycenaean Greece, home of the heroes of Homer's Iliad, a region covered by forests at that time. The forests provided materials for building and fuel (bronze workers and potters) and large tracts were felled, further and further from the main centres. Once again soil erosion became a major problem. Within 300 years the number of inhabitants in Greece fell by 75% owing to famine and impoverishment.

Stories like these, extending beyond forests, are pretty much a record of human development and expansion. They recur with each new occupation of a region by a new group of people: over-use of natural resources leading to declining agricultural production and supplies of timber and fish, followed by increasing human misery and migration. And today we're on our way to nine billion people with more than a billion of them hungry and malnourished, and there are no new places for migration.

The Global Footprint Network calculates Earth Overshoot Day,[6] the day when we have used all of what Earth can produce and regenerate in 1 year. It has moved from early October in 2000 to 1 August in 2017. So, for close to half the year we are living off our capital, drawing it down. Put another way, they calculate we need 1.6 Earths to supply what we consume and to regenerate what we use in a year. There are no unfished areas left in the open oceans and virtually all the fisheries are over-fished and in decline.

People with strongly vested interests in continuing as things are do not want to think about these kinds of facts (they are not alternative facts) and deny any need for major changes – there are indeed none so blind as those who will not see. But growing numbers of individuals, organisations and even some big corporations are increasingly uneasy about what they perceive as looming problems in the food–water–energy nexus. The trends in these three sectors, with their interactions and knock-on effects, have led many to the recognition that they cannot continue. The question is: will we humans stop them, or will the trends be brought to an end in catastrophic ways?

Trying to change the global trajectory to one that will allow the world to continue as we would like it to is no easy thing. From individual to global scales, we are all but locked in to the way we relate to and use natural resources. Although many leading economists are opposed to it, the current dominant, entrenched system of growth and its attendant behaviours ensures we keep growing at ever increasing rates.

But there is another way. Rather than short-term growth, it aims to build the resilience of the systems we depend on – natural and social, from local to global scales – with the aim of increasing human wellbeing without producing more than we need. In the face of the multiple looming threats and the uncertainty that surrounds how, where and when they may play out, the world needs a transformative change to this alternative, resilience-based future. A future that will enable us to absorb whatever shocks the interacting food, energy and water shortages and problems such as rising antibiotic resistance might bring.

What is resilience?

The word 'resilience' is now commonplace: in ecology, psychology, sociology, engineering, urban development, agriculture, industry, even as a personal philosophy. And in all of them it is used in somewhat different ways, thereby running the danger of becoming meaningless jargon. This would be most unfortunate because, if we are to successfully navigate the very real and dangerous global trends, it is resilience that needs to be understood and fostered. A particular concern is that, across all the areas in which it is used, some of its interpretations are just plain wrong, and if applied to a complex system such as the world or the local ones we all live in, they are misleading. In particular, three misconceptions stand out, as described below.

First, resilience is often referred to as the ability to 'bounce back', but it's not about bouncing back. It is the ability to absorb a disturbance and in the process re-organise so that the system (whatever it is) stays much the same kind of system: not exactly, but functioning in the same way, retaining its identity. It doesn't go back to just how it was. It 'learns' from the disturbance by altering the amounts of its different parts and the relations between them, and so makes it better able to deal with such a disturbance in the future. It happens in all kinds of systems and, in describing the myths of resilience in *9 Ways to a Resilient Child*, psychologist Justin Coulson likewise emphasises that it's not about bouncing back, it's the capacity to adapt successfully to disturbances.[7] Social systems learn instinctively and cognitively; natural systems learn by constantly changing through their responses to changes in their environment, captured beautifully in a poem by Jane Hirshfield:

> *More and more I have come to admire resilience.*
> *Not the simple resistance of a pillow, whose foam*
> *returns over and over to the same shape, but the sinuous*
> *tenacity of a tree: finding the light newly blocked on one side,*
> *it turns in another. A blind intelligence, true.*
> *But out of such persistence arose turtles, rivers,*
> *mitochondria, figs – all this resinous, unretractable earth.*[8]

The second misconception is the supposition that resilience is always a good thing. Resilience is neither good nor bad. There are lots of examples of very undesirable yet very resilient systems: inner city slums, landscapes that have become salinised, lakes that have turned into a kind of pea soup with algal blooms, evil dictatorships – the list goes on. The increasing use of 'resilience' as something that is always desirable misses the point that it is a property of a system and sometimes the need is to reduce resilience to get a desirable change.

Confusing resilience with resistance to change is the third misconception. The initial reaction of most people to disturbance and change is one of protection. But, in fact, change, and probing the boundaries of resilience, is necessary for maintaining and building resilience. Overly protecting a system, trying to prevent change and keep things constant, reduces resilience. A forest from which fire is always excluded eventually loses the species able to withstand fire: the only way for a forest to remain resilient to fire is for it to be burned every now and then. Children who are prevented from playing in dirt grow up with compromised immune systems. The only way to make children resilient to the environment around them is to expose them to it.

Overly protecting a system reduces its resilience, whereas probing its safe boundaries without crossing them builds resilience. It's not always clear how to do this, especially in humans, and Coulson describes how a 'suck it up princess' approach to children does not lead to increased resilience and can have very negative effects. In this case a helping hand to deal with disturbance is needed.

All complex systems have threshold effects: limits to their ability to adapt and keep going as they are. If the threshold is crossed, they begin to change in a different direction. In purely biophysical terms, your body, for example, is a self-organising complex system with limits to its resilience. You maintain a constant body temperature of 37°C. If it goes up you start to sweat, which evaporates causing your body to cool; if it goes down your muscles vibrate (shivering) so it goes up again. In technical terms, your body has negative (dampening) feedbacks to keep it functioning in the same way, but if the amount of change exceeds a certain limit, ~42°C on the upper side, the feedback is lost or switches to positive, it can't cope and you die.

There are many kinds of thresholds: how small an ecosystem can become before it loses species; how much runoff from agriculture can flow into a coral reef before the predominant species changes from coral to algae; the level of predation where the prey population changes from growing to declining; the amount of debt a company can manage before it heads into bankruptcy. But resilience is more than just staying away from thresholds. It's about developing the ability to change where a threshold occurs in order to become more resilient; to increase the 'safe operating space' of the system – especially in social-ecological systems. And crossing a threshold is not always a negative thing. Sometimes the system is in an unwanted state and the problem is not being able to cross into a desirable one. (Note: The 'state' of a system is determined by the things that define it: its identity. In the body temperature example, there are two alternative states: alive and dead. From the viewpoint of those to whom a particular ecosystem matters, its alternative states could be a healthy, productive one with lots of different species or a degraded one with bare soil, few species and unable to recover.)

In social systems, thresholds tend to be called tipping points. A trivial one is the take-off of fashion fads. More serious is a tipping point for riot behaviour in crowds. When something angers a crowd and they begin to riot, if the source of the anger goes away before some critical proportion of the crowd is rioting, the riot dies away. Beyond that critical level (seemingly around 20%), even if the source of anger goes away, the riot continues to spread. In business, the debt-to-income ratio and the number and diversity of suppliers have both been shown to have tipping points.

Resilience, then, is the capacity of an organism, an ecosystem, a business, a city, to absorb a disturbance by re-organising so as to keep functioning in the same kind of way and not cross into a different state of the system with a different kind of identity, or even into a different kind of system. In essence, it's about learning *how* to change in order not to *be* changed. Confronted with how to do this, how to build or manage resilience, it's not always easy to see how it applies in a particular situation, how everything fits together. What are the attributes of the system that determine its resilience? What thresholds are there and what determines where they occur? How can you make the system learn to stay away from them, or how can you get the system across a threshold? Because there will almost certainly be thresholds you don't know about, how can you make the system generally resilient to all kinds of disturbance? And, in general, how can this kind of understanding be incorporated into government and corporate behaviour?

This book is an unfolding story around these questions. The answers emerge from efforts to understand how natural systems and social systems work, and why it is so important to aim for a resilient world rather than try to keep it on one particular trajectory based on the impossible goal of never-ending growth: a trajectory fraught with increasing problems and negative consequences. A resilient approach on the other hand offers hope for a viable future, and having hope is in itself an attribute that confers resilience.

PART II

Encountering resilience in nature

3

Living together in ecosystems

'Form and function are one, joined in spiritual union' (Frank Lloyd Wright)

At the upper end of Botswana's Okavango Delta a narrow channel runs east for a few kilometres before spilling into the Savute marsh, which in fact can be either a marsh or an open grassland, depending on the levels of water in the delta. Mostly it's a mosaic of marshy patches and grassland, the grassy edges grading through increasing numbers and size of shrubs into open woodlands. It is a spectacular wildlife area, with seasonally migrating herds of wildebeests and zebras and an abundance and diversity of resident herbivores together supporting a rich array of predators. Particularly well known for its lion prides, it's a very good place to examine how predators and their prey get on together, which is what Petri Viljoen was trying to work out. In particular he wanted to determine what effects lions had, not only on their prey but on all the other species, both herbivores and other predators (i.e. the ecological role of lions).

'It's not just the animals that lions eat' he explained, putting forward his research proposal. 'It's also the effects the lions have on what other predators eat, especially hyenas. And the reverse effects of hyenas on what lions eat, because hyenas harass lions and can chase them off a kill.'

A lion pride makes a kill every few days and its behaviour follows a fairly typical pattern. After sleeping for most of the day, one of the adult females slowly gets up, yawns, stretches and begins to move. Sometimes the choice of direction is obvious – the prey are within sight or she can hear them. At other times there seems to be no good reason for the direction and, to establish their choice of prey, Petri needed to follow and observe them when hunting. For this he placed a radio-collar on one lion in each pride and on one of his tracking sessions he was aided by my ecology students on a field trip at Savute (well, actually, he was kind enough to let them tag along). The camp had its own light plane and I accompanied him on a preliminary reconnaissance flight to find the lions he wanted to follow.

'Which pride are you looking for?' I asked, pulling the protective thorn bushes away from the wheels of the plane – hyenas have a fondness for aeroplane tyres – while Petri carried out the pre-flight checks.

'I'd like to locate Scar-leg's pride again. They've been taking impala, tsessebe, warthog and buffalo, and now with more wildebeests and zebra they've had a few skirmishes recently with the pride from the southern side of the marsh. And the hyenas are making the most of it.' Scar-leg was one of six adult females in a pride that operated along the northern fringe of the marsh. Petri thought that, with the northern migration of the zebra and the wildebeests, the prides had been changing their boundaries and moving into contact more often. Was their choice of prey influencing, or being influenced by, this?

We took off and flew east along the northern fringe of the marsh while Petri tuned the receiver to the transmitter in Scar-leg's collar. He banked to the left at 500 feet above the ground and flew in a wide, slow circle. I looked out at the antennae attached to the wing struts and beyond. It was a beautiful, clear afternoon and, as we turned, the view graded from open grassland with groups of wildebeests and zebra into the mopane scrub where Petri suspected the pride might be. We had barely crossed over the edge of the scrub when he exclaimed 'Aha! I've picked her up'. He twiddled the receiver, banked again and a minute later smiled and pointed to a patch of scrub looming up in front of us. 'She's in there.'

We returned to camp and within an hour the whole group was on its way in a Unimog fitted out for fieldwork: a very large four-wheel drive vehicle with high clearance and high sides to the open back. We located the patch of scrub, which was actually quite open at close quarters, drove slowly in and sure enough there was Scar-leg and her pride: five other adult females, the pride male and several half-grown cubs. We took up a position some 50 m away. There was no need to worry about wind direction because the lions were accustomed to the vehicle and took no notice of it. Then came the boring part of lion research: waiting for something to happen. If the pride had recently fed, it was not uncommon on these vigils to wait all night while the lions continued sleeping.

This time we were lucky. After about an hour, one of the lionesses got up, looked around and slowly started moving. The other lionesses joined her and they began moving purposefully in a direction away from the Unimog. The male and the cubs stayed where they were. Petri started the engine and we moved off slowly in pursuit. So that we did not influence the hunt, we stayed well back, catching only intermittent glimpses of the lionesses up ahead. We followed them for a few hundred metres before they stopped, looking ahead of them. We waited for some minutes and then noticed that the male had come up behind us. He too stopped and lay down. The lionesses then split up but we were obliged to stay where we were because one of them was still in front of us. We strained our eyes and ears but could detect nothing. Suddenly, however, the remaining female leapt forward and the male also got up and started moving.

'There's been a kill' said Petri, starting the motor and moving after the male. A short distance further on, the bush thinned out grading into grassland and in the fading light we could see the lionesses on top of a young wildebeest. He was still alive making futile kicking gestures with his hind legs while the lioness that had brought him down kept her jaws firmly clamped on his throat. The other lionesses had meanwhile begun to lick him and tear open his stomach. The wildebeest gave a last heave and lay still. The lioness that had made the kill released her grip and moved to join her mates.

At this point the male arrived. He broke into an aggressive run, swiping and snarling at the females, which backed off while he lay down facing the stomach and began to tear at it and eat. The females approached again and joined in the feast. The cubs also tried to get into the act but were dealt with severely when they got in the way of an adult.

The popular image of the noble lion arises from the evocative, but often wildly incorrect, writings of early natural historians. In his *History of the Natural World,* first published in 1774,[9] Oliver Goldsmith, poet, natural philosopher and historian, describes the lion thus: 'To pride, courage, and strength, the lion joins greatness, clemency, and generosity; but the tiger is fierce without provocation, and cruel without necessity.' Given this was one of the first really popular books on natural history, it's not surprising the misconception took hold. It has been perpetuated in umpteen romantic novels, reaching a peak of inaccuracy in Disneyland's view of nature, *The Lion King*. It has nothing at all to do with the real behaviour of lions. Lions are not altruistic and noble. In reality they are opportunistic carnivores not averse to scavenging carcasses and that let their offspring die in times of hardship in favour of their own survival. Mothers love and lick their babies and they all get on fine when conditions are good, but it's the crunch times that define critical aspects of ecological behaviour.

It's not the strongest lions that survive such difficult times, but rather those with the physiological adaptations and behavioural traits that give them an edge over others. Over countless generations, survival and selection during such times has resulted in populations of lions that are able to cope with hardships. Darwin emphasised that evolution in species does not favour survival of the strongest, but rather those that are able to change and adapt; and it's the same for the evolution of resilience in communities – of animals, plants and people.

Within 15 minutes of the lions starting to feed, the first hyena arrived. It was getting dark by now and Petri brought out the spotlights, fitted with red filters so they did not blind the animals. More hyenas arrived, slouching and slinking around the carcass at a distance of ~10 m. We had heard their familiar prolonged 'yi-ip' calls earlier and now, close to the carcass, they were doing their characteristic cackling. Three or four of them edged towards the carcass, cackling loudly and harassing the lions, but a swift charge from the male sent them scattering. We stayed for another hour or so, by which time most of the lions had finished feeding and had lain down next to what remained of the wildebeest.

The hyenas were unsuccessful in their attempts to get a feed that night but this was not always the case. On this occasion a pride male had been present. In the absence of a pride male, even though other males may be present, hyenas are far bolder and can sometimes chase lions off a kill. A pride male has an aura about him: an 'X' factor. Virtually all cases of takeovers by hyenas of a lion kill that Petri observed involved prides with no pride male. So, despite his uncouth, chauvinistic behaviour, the male lion serves a useful function beyond that of stud. On several occasions Petri saw lions that had been chased off a carcass hunt and kill a second animal that same night.

This competition between lions and hyenas intensifies the predation effect of the lions, increasing the amount of carcasses for hyenas and all other scavengers, and so contributing to the welfare of these populations. But hyenas are not only scavengers. They are also predators, taking a smaller size of prey. The separate effects of lions and hyenas on the herbivore community would be different from what actually occurs as a result of their interactions. The interactions make lions

Hyenas taking over a lion kill in the Savute region of Botswana, forcing the lions to make another kill, probably a larger animal than hyenas could take on their own. In this case the hyenas were successful because there was no pride male present (photo: Petri Viljoen).

kill more of the larger herbivores than they would if left to their own devices, with the hyenas taking correspondingly fewer of the smaller ones.

The richness of all the predator–prey interactions in Savute makes detecting the effects of competition between predators difficult, but a very clear effect comes from a much simpler ecosystem in Western Australia. Several species have been driven to extinction there by introduced foxes and cats and now occur only on a few cat- and fox-free islands off the coast. Two of these species, burrowing bettongs and bandicoots, were selected for re-introduction to the mainland on a small fenced-off peninsula called Herrison Prong. It has a narrow neck connecting it to the rest of the mainland and so was easy to fence. A first requirement before the re-introduction could begin was to eliminate the foxes and cats, which were at high levels there thanks to introduced rabbits.

Foxes were trapped and eliminated first, despite determined efforts by those on the mainland side of the fence to re-invade. To thwart these efforts, the fence was extended out into the sea for about a hundred metres and late one evening Jeff Short, the scientist in charge of the project, saw a fox actually swimming along the fence, but it was too far and it turned back. After the foxes had been eliminated there was a dramatic increase in cats, with a corresponding decline in small native animals such as skinks and lizards. Cats were more adept than foxes in catching them. Competition between foxes and cats, and some predation by foxes on kittens, had controlled the effects of cat predation.

Unexpected outcomes like this, through indirect effects of a change in predators, are quite common. What happened after wolves were re-introduced into Yellowstone National Park in the USA is another example that took everyone by surprise. After wolves were eliminated in the 1920s, elk populations rose to high levels and their heavy browsing prevented regeneration of willow trees on the flats alongside rivers. Wolves were re-introduced in 1996 and by then these flat areas were open meadows. The numbers of elk killed and eaten by the wolves did indeed reduce the population of elks but nowhere near enough to significantly affect their browsing pressure. So what then came as a surprise was the emergence of lots of willows and aspens in several areas owing to huge reductions in elk numbers. These areas were later described as 'high-predation risk' areas, where escape from predators was difficult due to the terrain around the flat area adjacent to rivers. It didn't take the elks long to figure out that wolves were bad news and that going into some areas was just not a good idea and so they avoided them, allowing the willows and aspens to regenerate. This secondary effect was therefore a behavioural one. The presence of wolves significantly influenced the movements and spatial distribution of the elks far more than the overall number of elks.

An intriguing outcome of a different kind of behavioural response is the famous cycle of snowshoe hares and lynx in northern Canada. They show alternate highs and lows over an approximate 10-year cycle. How it happens was worked out by Charley Krebs and his colleagues, who have studied these two species for many years with a succession of students in the Kluane Lake area in the Yukon. For predators and prey to go through cycles of high and low numbers of each, there has to be a lag effect in response of one or both to the other. Without such a lagged response, there would be more or less constant numbers of each. Charley and his students found that the direct effect of predation by lynx on numbers of hares is not enough to cause the lag effect needed for a cycle. In a neat piece of research, they found that the lag effect was due to hormonally transmitted effects in female hares that had been scared half to death by being chased, but not killed. Female hares that survived this trauma had significantly reduced reproduction the following year, and this created the necessary lag effect that produces the cyclic pattern in lynx/hare populations. So if lynx were better, more efficient hunters and nabbed every hare they went for, the famous highs and lows of the lynx/hare cycle wouldn't occur. There would be constant, lower numbers of both, fluctuating somewhat with seasonal differences.

Continuing with the theme of cycles, as the forest industry in eastern Canada developed the foresters were puzzled by recurring outbreaks of a budworm in the spruce forests. It's a serious pest that eats the leaves of spruce trees to the point where the trees are killed. For what seemed at first to be no good reason, the populations of budworms suddenly explode every 40 years or so, killing the patch of forest they are in. The story was unravelled by Crawford (Buzz) Holling, the father of ecological resilience theory,[10] and included in his classic paper that triggered so many people (including me) to start examining resilience in the systems in which they were involved.[11]

Buzz found that when the forest is young and the total surface area of the leaves is small, the budworms are controlled by the birds. Because the birds can find them easily, the rate at which the budworm populations can increase is less than the rate at which the birds eat them, so they stay at low levels. When budworm numbers get very small, the time the birds have to spend searching for them gets too long so they turn their attention to other insects, which allows the budworms to start increasing until they once again attract the attention of the birds. They are therefore kept in what's called a low-density 'predator pit' in which the population shifts up and down around some fairly low equilibrium amount. But, as the forest grows, leafiness increases, and the budworms are progressively harder for the birds to find, and eventually a threshold point (tipping point), is reached where the rate of budworm increase is higher than the rate of consumption by birds, and the population explodes to where the only thing that can stop it is lack of food. The trees are defoliated and die and the forest is taken over by birch trees. Spruce trees are superior competitors and so eventually they come back. This can take ~50–100 years and then the cycle begins again.

Virtually all well-studied predator–prey examples reveal complex, non-linear dynamics. Some exhibit regular tipping point effects, others involve complex secondary and feedback effects that keep them and the species in them resilient to environmental and other changes. Trying to manage such systems requires knowing where and when, or where and when not, to intervene in these dynamics.

Pyramids and webs

In order not to lose weight, an average cow on a rangeland has to eat ~4000 kg of grass a year. And, in order to supply it, the grass layer has to produce many times that to keep itself going and even more to reproduce itself. The result is the first step in what's known as a food pyramid, because the same process occurs up the food chain. In natural ecosystems, when an animal gets eaten and converted into another animal, energy is lost in the process, and animals can only eat a portion of what makes up their food supply. Enough prey must be left to reproduce themselves.

The minimum number of levels for a functioning food pyramid, and therefore functioning ecosystem, is two – almost invariably plants and decomposers that recycle the nutrients in dead plants back into the soil. But there's an interesting variant in Namibia that highlights the way food webs and pyramids work.

The Namib Desert along the south-west coast of Africa is one of the driest regions on Earth, with some truly spectacular sand dunes reaching heights of more than 250 m. They extend inland for over 100 km and their northerly extent reaches the Kuiseb River, which marks the boundary between the dunes and the gravel plains. The Kuiseb rises in the Khomas Hochland (high plateau) near Windhoek, Namibia's capital city, and flows west across the desert to the coast at Walvis Bay. About 50 km inland, at a place called Gobabeb, is the Desert Ecological Research Unit built on the northern bank of the river. In the 1970s and '80s, its director Mary Seely maintained a program on how desert ecosystems work, including some pioneering research on amazing beetles that inhabit the dunes. They are part of the dune food pyramid and the way they get their water is fascinating.

One of the striking features of the Namib is its fog. Cold water wells up against the African continent from deep in the Atlantic Ocean, flowing northwards in the Benguela current. Easterly air is rapidly cooled as it passes over this cold water and dense fog banks roll in from the sea. Much feared by mariners, the combination of impenetrable fog and treacherous sand banks has claimed the lives of many ships and their crews, giving rise to the name Skeleton Coast.

The fog travels all the way to Gobabeb and beyond, and for years at a time it is the only moisture in the desert. Not surprisingly, organisms have evolved to take advantage of it. Being at Gobabeb when a fog happens offers a unique experience. A fog warning device goes off, invariably in the wee small hours, and very quickly you are up and off in a fat-tyred four-wheel drive vehicle. With headlights piercing the flowing mists, the vehicle crosses the riverbed into the dune fields. Cold wet air swirls through the windows and the world seems to have closed in.

The vehicle stops at the base of the study dune and as the engine dies you are enveloped by silence. You scramble to the top of the dune, vision limited in the flashlights to a metre or so around you. It is difficult to orientate oneself. Arriving at the knife edge crest you lie with your head just above it, look around to gain your bearings in the dim grey light, and are confronted by one of nature's truly amazing spectacles. Lined up along the crest are dozens of beetles, heads down, standing on their front legs with their backs facing seawards into the rolling fog. They stand motionless as the fog swirls past them. As it encounters their backs it condenses and runs down to collect in small droplets at the tips of their snouts, and they slowly drink the water so provided.

The beetles aren't the only animals that have evolved to get their water from the fog. A small web-footed gecko lives underground during the day and runs around at night. When the early morning fog comes, the moisture condenses on its large eyeballs and it uses its long tongue to lick the water off.

How evolution came up with these behaviours is intriguing, but ecologically what is really interesting about this dune ecosystem is that it has no live plants, no primary producers. Yet it has a rich animal life, including these remarkable beetles and geckos. The base of the food chain is dead plant matter: bits of litter blown in

A Namib Desert sand dune with the dune beetle on its front legs, the shovel-snouted lizard and the golden mole (photos: Gobabeb, Mary Seeley).

from the inter-dune valleys. Because of the nature of wind patterns, these bits of dead grass collect in eddies on the lee sides of the dunes where they accumulate and become buried by the constantly moving sand.

Tiny invertebrates and fungi feed on the buried detritus and form the food base for a variety of very small animals such as mites and nematodes, which in turn provide sustenance for an array of slightly larger ones. Eventually, in the progression up the food pyramid, the animals reach a size we can see. There are by now far fewer individuals and among them are Mary's beetles and the animals that feed on them, including strange specialised lizards (the shovel-snouted lizard) that perform a kind of dance on the scorching sands by standing on two feet at a time, opposite front and back, while they allow the other two to cool off. They in turn are fed upon by a blind golden mole that 'swims' under the sand. It detects the presence of life on the surface by vibrations and emerges in a flurry of activity, like a killer whale rising out of the sea, disappearing again in seconds with its prey.

These fascinating dune ecosystems have developed over many thousands of years. They are simple when compared with tropical rainforests, but perhaps more impressive by virtue of the fact that the entire food pyramid, with its web of highly specialised species, has been maintained for all this time without any plants of its own, relying solely on a supply of bits of dead plant matter blown in from elsewhere.

Scores of food pyramids have been described in all kinds of ecosystems and those with vegetation all show the same basic features: a relatively large mass of plants with a much smaller mass of herbivores eating them, and a smaller mass of predators eating the herbivores. But ecosystems are more complex than just this pyramid shape of biomass. Different kinds of herbivores prefer different kinds of plants and different kinds and sizes of predators eat different kinds and sizes of prey. Furthermore, it's tricky to proportionally allocate omnivores – such as baboons, mongoose, bat-eared foxes and us humans that eat both plants and animals – to different layers. All in all it becomes a food web in which each layer in the pyramid has a lot of different kinds of species with lots of connections between them, and also who-eats-whom connections with the species in the layers above and below. The many connections in diverse food webs allow for the whole system to keep going if a particular species is lost.

There are many interactions between species that make for a complex food web, such as the competition between different predators for the species of herbivores they prefer, which in turn eat, and thereby change, the amounts of the different species of plants. And, in addition to the more catholic tastes of omnivores, there are scavengers eating dead animals, and all of the dead remains and animal faeces get eaten by a host of underground 'reducers' that leave the remains of the dead stuff they eat in tiny pieces that are used by the decomposers (fungi, bacteria) that recycle the nutrients for plants.

Much of what an animal eats simply goes straight through it. The kind of digestive system it has determines how long this takes, the kind of food it chooses, and how efficient it is in converting its food into its own body. Herbivores are the least efficient and as the quality of the food declines the proportion passing through increases. Elephants are probably the least efficient of herbivores and in times of drought they have to eat even more than usual to keep going, so even more passes through them. At times they have to subsist on just dead leaves and bark from trees, and it's thanks to this that I happen to be the owner of the biggest piece of dung known in the world. I was driving through Botswana's Tuli Block in the 1982 drought with Bruce Page, who was doing his PhD there, when I spotted this specimen. We stopped and got out the Land Rover to look down in awe at a piece of dung 60 cm in length with a circumference of 61 cm. It was in fact four boluses (a ball of elephant dung is called a bolus) entwined together by mopane bark. Elephants produce on average one bolus every hour. In this case, the string-like inner bark had prevented separation of the mass of partially digested material and four boluses had remained firmly joined together. We stood there imagining the discomfort it must have caused its owner and the relief when it was finally expelled. If elephants can smile it would surely have done so. I took the dung back with me, soaked it in polyurethane, and it stands to this day against the fire-place in our sitting-room.

Bruce Page admiring the largest known piece of dung in the world, produced by an elephant that had to eat the bark of trees during a drought in northern Botswana. Elephants have one of the lowest rates of converting food into body mass (photo: Brian Walker).

As it progresses up the food chain, the loss in efficiency of conversion contributes to the pyramid effect but the levels of inefficiency are not nearly enough to account for the huge differences in biomass between the layers that the actual pyramid displays. The whole system seems to be so inefficient. If you could increase the efficiency with which each layer eats the one below you could have many more big herbivores and predators. Why is there so much 'wastage'?

The Savute ecosystem that Petri studied reveals how big the differences are between herbivores and predators. His painstaking measurements showed that the ratio of the total weight of all the lions to the total weight of all the herbivores ranged from 1:52 in the dry season to 1:72 in the wet season (when the zebra herds moved in). In other words, the total weight of all the prey was always at least 50 times more than the total weight of lions. The total weight of hyenas is about the same as that of the lions (there are more of them, but they are smaller) and so on average the ratio of the weight of large predators to the total weight of large herbivores at Savute is around 1:25, so it takes around 25 units of prey to support a unit of predator. In order to sustain themselves, the 87 lions of Savute made an average of 1011 kills a year, made up of 303 buffalo, 242 warthog, 164 zebra, 79 tsessebe, 49 wildebeest and sundry other minor species including impala and even porcupines. It is a very complex food web with big differences in biomass between the layers in the food pyramid.

It is the apparent inefficiency in food pyramids and the wide choice of food for predators and herbivores in the food web that allows them to absorb and survive shocks like droughts, fires and diseases. It isn't wastage: it's what makes them resilient.

Cascades, keystones and genes

The most popular target for anglers in the lakes of Wisconsin is the largemouth bass. However, it isn't a preferred fish for the table (at least in Wisconsin) so it's mostly catch-and-release recreational fishing, which means the population of largemouth bass remains at high levels. This is significant because it is the top predator in these lakes. Its preferred prey are small-bodied minnows such as golden shiner, fathead minnow and various species of dace, and with a large population of the bass this second layer of fish are kept a reduced levels. They in turn feed on a variety of small zooplankton, animals that eat the even tinier phytoplankton (free-living algae) at the bottom of the lake food web, and that are necessary for the existence of all the other species in the lake. So, the more largemouth bass there are, the fewer algae there are.

This top-down process of the effects from predators through all the layers in the food web is known as a trophic cascade, and Steve Carpenter has been studying them in the lakes of Wisconsin for over 30 years. He has found that the trophic cascade process significantly affects the outcome of lake pollution.

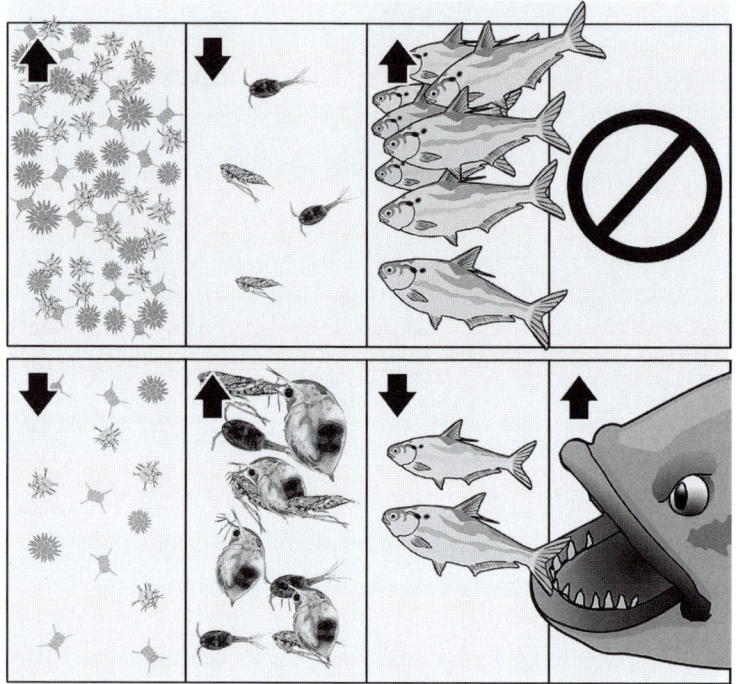

Cascade effect of having or not having (top row) a top predator in a lake ecosystem. Too much fishing for big predator fish in Wisconsin lakes has resulted in unwanted blooms of algae (illustration: Anthony Thorpe, Lakes of Missouri Volunteer Program, University of Missouri).

If nutrients in the lake are initially at low levels, any increase (e.g. through runoff into the lake from agriculture) that results in increased growth of the algae is simply translated up through the food chain, increasing the amounts of all levels – and the lake becomes more productive. But if the nutrients continue to increase there comes a point where the amounts and growth rates of the algae swamp the zooplankton, resulting in algal blooms that make the lake more like pea soup than the clear lake it once was. The level of nutrients where this tipping point happens is quite high when there are lots of bass, and therefore lots of zooplankton. But if bass numbers are reduced the tipping point occurs at a much lower level of nutrients. So reducing the top predator in the lake makes it more susceptible, less resilient, to nutrient pollution.

The wolves to elks to willows process, described earlier, is a terrestrial example of a trophic cascade, initiated by re-introducing the predators. In contrast, an example of a disrupted cascade is currently unfolding in coral reefs around the world. The headline news about coral reefs is the massive coral bleaching events, and we'll come back to them later, but our interest here is on these trophic ecological effects, independently of bleaching.

There are over 2500 species of corals, and their variety, shapes, colours and beauty, and the many colourful fish they support (like the 'Nemos' of the reefs), is wonderful to see. But corals are declining and even disappearing on many reefs. Terry Hughes, Director of Australia's Coral Reefs Centre of Excellence, has been studying and recording this decline over some decades, especially in the Caribbean and Australia's Great Barrier Reef. What he has found is that many of the corals in the Caribbean have almost gone and parts of the Great Barrier Reef are beginning to struggle.

On healthy reefs, big predator fish keep smaller ones at lower levels, so the effect of these smaller ones on the layer below them is reduced, and so on down. The net effect of the trophic cascade is to allow high numbers of the herbivorous fish that eat the filamentous algae that grow on reefs, and that compete with corals for space. As fishing pressure on big fish increased, the cascade led down to fewer fish that eat algae, giving the algae a competitive edge. And now the algae-eating fish are also being harvested, and algae are replacing corals. In much of the Caribbean, what used to be splendid coral reefs are now beds of filamentous algae, with very few live corals.

Top predators increase the diversity of all the other species in an ecosystem: on land, in lakes and in the ocean. Continuing with the story of re-introducing wolves into Yellowstone, the movement of elks and re-establishment of willows and aspen was the initial effect, but, as time went on, further secondary effects emerged. The willows allowed beavers back in and they and their dams increased in numbers, changing the hydrology of the rivers and the species in them. Other shrubby plants came in under the willows and aspens, producing fruits that attracted back bird species, and further such knock-on effects increased the diversity of the whole ecosystem. When top predators are removed or greatly reduced, the variety of all the other species is reduced. The ecosystem becomes much simpler – and less resilient to disturbances of all kinds.

Somewhat different to top predators, but equally important in their effects, are 'keystone species'. They can be at any level in a food web and, as their name suggests, if they are removed it leads to knock-on effects that radically alter the whole ecosystem. A humble banksia plant in Western Australia has become a keystone species as a result of land clearing for wheat production. The region is renowned for its diversity of banksia species and their flowers attract a rich variety of honeyeater species that pollinate the flowers as they sip their nectar. The co-evolution of the array of banksias and honeyeater species has ensured there is a continuous supply of nectar for the honeyeaters, thanks to a sequence of flowering across all the banksia species. At any one time, there are several species in flower and the honeyeaters visit and pollinate them all while getting the nectar they rely on.

Only 3% of the huge wheatbelt is still natural vegetation, in the form of thousands of isolated fragments and, at one particular time of the year, there is now just a single banksia, *Banksia prionotes*, producing nectar for birds. If for any reason it should disappear, there would be a fatal gap in the supply of nectar and all the honeyeaters would disappear. With the disappearance of the honeyeaters there would be no more pollination of all the remaining banksias and so eventually all of them would disappear. The fact that *Banksia prionotes* has become a keystone species was revealed only after a very nice study of the feeding behaviour of the honeyeaters in the region. Once known, it became a *focal species* for conservation, and this is in fact what Rob Lambeck (the conservation scientist who conducted the study), called it in the paper he published.

The dependence of the honeyeaters on this single banksia species illustrates an important element of resilience. As the number of banksia plants declines, the honeyeaters have a progressively harder time getting enough nectar to keep themselves going and, below some critical number of plants (before they are all gone), the honeyeaters will disappear. That density of banksias is a threshold level in their population that marks the change from a growing honeyeater population to a declining one, signalling their inevitable demise. As the number of banksia plants declines towards that threshold, the resilience of the honeyeater populations decreases: that is, it will take a smaller and smaller shock or disturbance (climate, disease) to the honeyeaters for them to be eliminated.

Well before the honeyeaters reach this threshold for minimum population numbers, their genetic diversity begins to decline and it too reduces the resilience of the birds, and it also has a threshold effect. Loss of genetic variability means loss of particular traits for adapting to the environment and for coping with shocks, and there is a minimum population size below which there is a progressive decline in these adaptive genetic traits and an accumulation of maladaptive traits. Gene changes are occurring in all species all the time; there is a continuing process of mutation going on, producing new forms of genes. The majority of these new forms have negative effects and in thriving populations they are quickly purged because the individuals that have them are unsuccessful. But in very small populations the individuals can survive and the genes can accumulate and become fixed; the process is known as 'mutational meltdown'.

Land clearing doesn't only lead to a decline in the numbers of animals and plants in a population, it also leads to loss of connectivity between populations, and this too can have genetic effects. Species with populations that have been completely separated from each other for a very long time have evolved the kind of genetics and population dynamics that make small populations viable, so changes in their connectivity isn't a problem. But it is a problem for those that have been able to occasionally interbreed, and the clearing of rainforest in northern Queensland has had this effect. The remaining patches of forest – on mountains that are quite far apart – are home to several birds that can only survive in a forest and, before clearing, the areas between the mountain patches had bits of forest and

Golden bowerbird male arranging his bower to attract a female. He sings loudly to proclaim his territory and to attract females. Slight differences in bandwidth and peaks in songs can switch off recognition by other males and females (photo: David Westcott).

woodland that provided a loose connection (a kind of corridor) for the birds. The connections are now broken and, in trying to determine the effects of this, David Westcott set out to see if birds of the same species in different patches were still able to interbreed. He chose the golden bowerbird to start with and recorded the advertisement songs of the males in their territories. When he played these songs in the territories of other males in the same forest patch, they attacked the microphone vigorously. But when he played them in the territory of a male in a patch on a different mountain it ignored the song. David then analysed the sound waves and they showed that quite minor differences in bandwidth and the number of peaks were enough to break the recognition – and therefore mate recognition and breeding. There is a critical amount of time that populations such as the bowerbirds can be completely separated before bandwidth and peaks are such that mate recognition is lost and they can no longer breed. Song diversity has arisen rapidly since the forests connecting Mt Baldy and Mt Edith were cleared. So time can be a threshold. Without conducting an unacceptable experiment in these dwindling forest areas, we will probably never know what the threshold amount of separation time is.

Having loosely connected sub-systems is an attribute of resilient systems. Years ago a friend and I were pulling together a set of ecological principles that managers and planners should bear in mind, and I asked several eminent ecologists to give me their top principle. When I asked the Director of Imperial College's Silwood Park, Sir Richard Southwood, he thought for a while and replied: 'It's the significance, the importance, of the difference between very, very small and zero.' Very small and infrequent connections are hard to identify and there are many of which we are and will remain unaware. And, as with the banksia species in Western Australia, it's very hard to detect or know about crucial keystone species. We should, however, be aware that when we change landscapes – and in fact when we make any changes in the structure of a system – there is a likelihood that we are disrupting loosely connected sub-systems and changing the ecological status of species, and we should be looking for signs of this.

'Form and function are one, joined in a spiritual union,' said the architect Frank Lloyd Wright. As is well known in the world of architecture, he was refuting an earlier saying, wrongly attributed to him, that 'form follows function'. One doesn't follow the other, said Frank. They are intimately linked and equally influence each other. In ecosystems, structure and function have co-evolved over a long time and if you change structure, function changes in response, and the changes in function then feed back to cause further changes in structure. The pattern of these reciprocal changes is usually of a dampening, self-correcting kind and the system returns towards something like it was before the first change. But sometimes, after a big initial change, the changes that follow are not self-correcting and instead cause the system to keep changing away from what it was, leading to it becoming a quite different system – such as the shift in Steve Carpenter's lakes from clear to murky green, or the change from increasing genetic diversity to mutational meltdown. These kinds of threshold effects are a cornerstone of resilience.

4

Ecological choreography

'Imagine how much harder physics would be if electrons had feelings!'
(Richard Feynman)

Choosing what to eat

How much do animals share or compete for food? The theory that arose around this question suggested that the diets of all the herbivores that live together in an ecosystem (their 'dietary niches') were largely separate from each other. Having little overlap means efficient use of all the plants and so, overall, more animals are possible. Evolution may have come up with this arrangement if it's assumed the selection pressure was for squeezing in as many animal species as possible: the common assumption in the 1950s and '60s. But several field ecologists had begun to question this. Tony Ferrar, a wildlife ecologist in Zimbabwe, was one of them. A careful observer, he had been thinking about how much separation and overlap there was in the diets among all the different herbivores he had been watching, and he had doubts about the niche separation hypothesis. So he set out to test it.

Tony began his career as a research officer responsible for game management surveys, starting in the Gona-re-Zhou game reserve in south-east Zimbabwe. He was investigating elephant feeding impact, which naturally included his interest in what all animals choose to eat. One Sunday afternoon he had gone out with his friend Cynthia to watch and photograph elephants feeding. Just before sunset they stopped in a patch of scrub to watch a lone cow with a 4-year-old calf. Through binoculars Tony was watching the selection of shrubs. It was basically a patch of mopane scrub but among these there were some other species and the cow seemed to be selecting one of them. Then he noticed she looked pregnant while still having a calf in tow, although with all the scrub it was difficult to be sure.

'Hmm. That's a bit odd' he said, turning to Cynthia. 'I think I'll get a bit closer for a better look. I need some close up photos of elephants with calves, anyway. I won't be long. You stay here'.

He got out the Land Rover and started walking towards the elephants. As he did so, his dog, McGrooch, jumped out and followed him. At a safe distance Tony stopped but McGrooch took it into his head to rush up to them barking. The cow charged him and he promptly ran for shelter behind Tony. Having no other option Tony turned and ran for the nearest tall trees, but tripped over one of many coppicing stumps. While still bent over, the elephant struck him between the legs with one of her tusks, knocking him head over heels so that he landed on his back with his head towards her.

She bore down on him piercing his lower abdomen with a tusk, then lifted her head and crushed his pelvis with her knee. Tony held on to her foot with both hands above his chest, staring at her toenails as he heaved upwards to get rid of her. The futility of trying to bench press an elephant crossed his mind as he lay there screaming at it. He has a vivid memory of looking at the setting sun between her breasts (elephants have twin breasts between their front legs, shaped and located as in humans). As she lifted her knee he rolled out, pushed himself to his feet and tried to run but discovered his legs wouldn't work properly. His pelvis had been broken in four places. His lurching attempt, however, carried him a few metres and he fell into a thick patch of scrub. The elephant turned around, trumpeting fiercely, swung her head back and forth a few times and charged off.

As the sounds died away and Tony did not return, Cynthia plucked up courage and made her way cautiously towards where he had gone, calling out to him. He called back and when she found him told her how to get back to the Park headquarters, knowing his injuries were such that he shouldn't move. The warden at HQ radioed for help, drove out to Tony and lit a large signal fire, and late that night he was helicoptered out. After a few weeks in hospital he made a remarkable recovery with virtually no after-effects.

Tony was later transferred to Lake Kyle National Park and started to investigate more closely how all the different herbivores in that ecosystem divided up the plants on which they depended. The 'niche separation' theory implied that whatever happened to one herbivore species would have little or no effect on others since their feeding niches were separate. Although this might be about right for extremes such as zebras and giraffes, Tony doubted its generality and tested the theory by measuring the feeding niches of the 14 large herbivores in Kyle. In each of the different habitats he carefully recorded all the different species of plants each herbivore ate and how often they chose to eat them. His findings were at odds with the theory of non-overlapping niches. After one of my visits to his observation sites he showed me the results he was getting.

'What these data suggest' he said, running his finger over a series of intersecting graphs, 'is that there is clearly much more overlap than separation. Sure, pure browsers don't overlap with pure grazers but apart from that there is a lot of overlap. In fact much more than I originally suspected.'

His data showed that impalas are the real generalists, browsing and grazing across the whole spectrum of woody, herbaceous and grassy plants. Among the grazers and among the browsers the selection was more subtle.

'We all know wildebeests select finer grass than buffalos, as the shapes of their mouths and teeth suggest. But take a look at all the grass species they've both been eating through the year and there is clearly a lot of overlap. They've been choosing different species at different times and, not only that, wildebeests start eating the new leaves of coarse grasses after buffalos have chomped them down. It's really hard to find any clear pattern in these graphs.'

There has been a lot research on what animals choose to eat, giving rise to a body of work known as 'optimal foraging theory'. A paper by a colleague of Tony's has the title 'What should a clever ungulate eat?' and it comes up with a very detailed model of how the dimensions and kind of digestive system of the animal match the combination of nutrients and textures best suited to it, and hence the kinds of plants it should select – its optimal diet. But it too has problems in matching the predictions from the model with field observations. What is best for each individual species (its optimal diet) leaves the whole ecosystem less able to keep going if one animal species or one kind of plant is lost. The plant–herbivore system is more resilient if each individual species has a wider than optimal diet, and over time – a very long time – this is what has selected the kinds and pattern of foraging behaviour in the community of herbivores.

The strict niche separation model was wrong. The reality of wide ranges and overlapping diets is a much more robust arrangement, conferring a high degree of resilience on the plant–herbivore system. If a particular plant or animal species is much reduced or eliminated, other species compensate, filling the gap, and the ecosystem continues to function in much the same way.

The 'overlapping niche' story strengthens the emerging understanding that variability, overlap, uncertainty and apparent inefficiency in the way ecosystems are put together, and in the way they function, allow them to persist in the face of environmental stresses and changes.

Choosing where to eat

The rhino lay dozing in the riparian thicket on a still humid morning, only its torn ears alive to the sounds of the bush around it; nothing detected, all was safe and well. Not far back along the path, Pete Goodman also tested the air: yes, he was still downwind from where he knew the rhino must be so he moved forward cautiously, peering through the scrub. He needed to see those torn ears so he could identify the rhino. They were like a thumbprint, a personal record of that rhino, and Pete had hand-sketched records of all the rhinos' ears in the Mkuze Game Reserve, which lies in the heart of Maputaland, in the northern part of Zululand. He was assessing the numbers and distributions of all the herbivores in the reserve so he could determine how the different kinds of habitats, and the condition they were in, influenced the diversity and distribution of the animals.

The questions Pete was asking were related to those Tony Ferrar had posed, but his focus was on the habitats the animals used and how they moved between them, rather than the particulars of what plants they ate. He was trying to come up with clear-cut relationships that would help the Reserve managers determine their priorities. Which kinds of habitat are particularly important for different animal species, especially the rare ones? Should they burn or not burn a particular area? Or put in or take out a water point?

Dense thickets along the Mkuze River give way to open plains and savannas that in turn give way to rocky hill slopes, and in each there is a variety of different kinds of plant communities. It's a good place to investigate the relationships between a diverse set of habitats and a diversity of animals. Pete needed to establish the numbers of different herbivores in each kind of plant community, which meant having to count black rhinos in the riverine thickets. The black, or more correctly hook-lipped, rhino favours these dense thickets, eating woody twigs and leaves. It is a far more aggressive and irascible customer than its grass-eating cousin, the white (square-lipped) rhino.

Pete inched forward and raised his head to get a better view and as he did so he detected wind on the back of his neck. The wind had changed. The rhino raised its head and sniffed the air: it detected trouble, rose and charged at the source of the smell in a lightning fast move. The last thing Pete remembers is thinking 'gotta get off the path'.

The rhino's horn hit him just below the chest, knocking him out cold. The game guards at the edge of the thicket saw Pete being tossed up and disappearing again before one of them managed to get a shot over the rhino's head and it charged off. When they got to him Pete was unconscious and bleeding profusely from holes in his chest and right leg. Blood was pouring down his head from a huge gash in his scalp, which had split when the rhino had thrown him against a tree. The guards staunched the bleeding and radioed for a helicopter, which eventually arrived and flew him to the Mkuze airstrip where a doctor and small plane were by now waiting, and he was flown to a hospital. The rhino's horn had stopped just half a centimetre from Pete's heart. The surgeons stitched up his perforated liver, his diaphragm, his stomach, two gaping wounds in his leg and the split in his head, and they set several broken ribs. Amazingly, he recovered with no permanent disability and within a few weeks was back analysing his game count data.

The welfare of the rhinos turned out to be closely tied to the density of the thicket vegetation. They did spend time in more open savanna as they moved between their favoured thicket areas, but there seemed to be a critical density or thickness of vegetation below which black rhinos were largely absent. What kinds of plant species were in the thickets mattered less. And when he considered all the animals – from the cryptic little suni antelope to the wildebeests and zebras to rhinos, kudus and giraffes – it turned out that their distribution was related much more to vegetation structure than plant species. But it was hard to tell just how vegetation and animal species were related.

There was a lot of overlap in what kinds of vegetation the different animals preferred, and a lot of variation between the same kind of habitat in different parts of the reserve and at different times of the year. In the dry season animal diversity was highest where there was most vegetation. In the wet season it was mostly the other way around, because the grass became too thick and the shrub layer too dense for selective feeders, but other animals still liked it. Sometimes the choice of place to forage appeared to be due to other animals, either deliberately foraging together or avoiding each other. A positive outcome of hanging out together in a larger group is spreading predator risk, but avoiding other species reduces competition for food. Sometimes it seemed to be just a chance outcome of contact with other animals. The lack of clear-cut relationships in the masses of data he had collected bothered Pete because ecological theory at the time led to expectations of strong relationships.

Fuzziness, chance and resilience

Apart from the similarity in their hair-raising field experiences, the similarity in lack of clear-cut relationships in Tony's and Pete's studies highlights a very important aspect of how natural ecosystems keep functioning through all kinds of disturbances. The effects are spread among all the different species through their overlapping diets and choice of habitats, with lots of variation in both through the course of the year as the seasons change. This is how these ecosystems have evolved; it is how they are able to persist. It is the very lack of clear-cut one-on-one relationships – the overlapping fuzziness and uncertainty in the way they are structured and in how the different plants and animals relate to each other – that makes them and the communities of plants and animals resilient.

The opening quote of this chapter, attributed to the famous physicist Richard Feynman ('Imagine how much harder physics would be if electrons had feelings!'), beautifully captures the fact that ecosystems just don't behave like physical systems. The organisms in an ecosystem have complex behavioural responses to the environment and to each other, and so they are always changing, often unpredictably. And this continual change is not just characteristic of them: it is essential for their continued wellbeing and for keeping the ecosystems in which they live functioning in the same kind of way. In physical systems the variance is regarded as background 'noise', and physics envy – the quest for elegant equations to describe ecosystem dynamics – can get in the way of seeking ecological understanding. What's noise to the physicist is music to the ecologist.

5

Disturbance, change and diversity

The architects of pattern

Morvan, Patrick and I were on a sampling trip taking measurements of the vegetation and soils in the Hwange National Park in Zimbabwe. They are two wonderful technical assistants who know all the plants and they were helping me to help the Park managers gather the information needed to answer their over-arching question: 'How do we keep the Park in its natural state?' This immediately raises the questions: 'What is its natural state? How many kinds of ecosystems are there are in Hwange? Are any of them changing? If so, should something be done about it?' Basically, the questions facing national park managers everywhere.

Most of Hwange is in the Kalahari Sands region of southern Africa and, where rainfall is high enough, they are clothed with forests of deep-rooting Rhodesian teak trees, their canopies covered with attractive purple flowers in the rainy season. There are also shrublands and grassy drainage channels fringed by woodlands of magnificent acacia trees, their big grey pods eagerly sought after by elephants. All in all, there are scores of species growing in a patchy mosaic of forest, woodland, scrub and grassland, supporting a wide variety of animals.

Before the Park was created in 1928, any piece of forest within ~15 km of the railway along its northern border could have been logged; beyond that it wasn't worth the effort. So we needed to find areas undisturbed by logging to compare with those that had been. We located several teak sites on aerial photos beyond 15 km and headed for one near a small 'pan' (a shallow depression filled with water in the rainy season) called Mtswiri Pan. It took its name from the 'Ndebele word for a species of tree: a large old one in this case, near the edge of the pan.

After sampling a couple of sites along the way we arrived at the pan late in the afternoon. The bare zone around it was covered in dung and despite some early rains it was still only half full. The water was green with a strong smell and quite saline, coming from the borehole adjacent to the pan, which was not working at the time. One would have had to be very thirsty to drink it. I drew up under the fine old mtswiri tree at the edge of the bare zone and laid out my bed roll alongside the Land Rover. Morvan and Patrick had surveyed the evidence of much recent elephant activity and opted to spend the night in the corrugated iron hut that housed the borehole pump, surrounded by a ditch designed to protect it from inquisitive elephants. The stifling atmosphere in the hut made the chance of elephants seem a lesser problem to me and, besides, it could barely hold two.

Evening came swiftly and we lit a fire, cooked and ate our supper and sat chatting for a while over mugs of tea. Morvan had been wondering about the pan as he looked at it across the fire.

'How does the water stay in it?' he asked, contemplating the still surface. 'If this is all sand why doesn't it just sink down?'

'Two reasons' I replied. 'You need a slight depression like this, and elephants'.

'Ah, yes,' commented Patrick, 'that was Dr Weir's work, wasn't it?'

John Weir had worked in Hwange a few years earlier and had found that, by splashing and trampling around, elephants puddled the surface under patches of shallow water that originally did seep down. With small amounts of fine soil that washed into the hollow, the trampling created a clay seal that got harder and harder and in the end a pan developed. Morvan turned and looked reflectively at the teak scrub, marking the edge of the sand just beyond the mtswiri tree.

'When you dig in that soft sand it's hard to believe there's any clay at all' he murmured. The amount of clay is in fact very small. But it's enough to do the job if it gets concentrated in a small patch. Where water collects in a depression, the elephants just engineer things and make the depression bigger and more impermeable. So often, animals are the engineers of landscape architecture.

Not long after dark we made ready for bed. I built up the fire and lay down between it and the Land Rover. I nodded off quickly but sometime around midnight woke to a soft sound. The moon had risen and I seemed to be looking at a forest of legs. A herd of elephants had arrived silently and were making their way to the pan. They had come from the direction behind the Land Rover and were peeling around it, staying away from the smoking embers in the fire so I was fairly safe, but I rolled under the Land Rover as more of them came down. I knew they had detected me as the ends of their trunks appeared periodically under the upslope side of the Land Rover, and I could imagine them giving the vehicle the once over. I felt no sense of danger and the elephants were completely quiet – vocally, that is. Their constant stomach rumbles were something else, punctuated every so often by a stupendous fart. After a while they had all passed by and I could hear splashing down at the pan, ~20 m away.

I rolled out from under the Land Rover and climbed onto the roof. I could see the elephants quite clearly in the moonlight as they moved around the pan, sucking up water in their trunks, folding them back into their mouths and squirting it down their throats. There were just on 30 of them. Periodically some would raise their trunks and turn them towards me, like periscopes in the moonlight. It was a still night and I couldn't detect any wind but their senses were much sharper, and the times when they pointed their trunks towards me coincided with slight shifts in the movement of air. After about a half an hour they moved off, drifting into the trees on the far side of the pan, not back to where they had come from.

The herd was moving south and had stopped for a drink en route. With the rainy season underway, they were heading for the normally drier southern areas, which would have had a long rest from grazing and browsing and now, with seasonally available water, would be an attractive source of good and different food. This would also allow the dry season feeding areas they were leaving to recover. The two different kinds of area allowed the system as a whole to have periods of growth and recovery: a more resilient arrangement than one with constant herbivory.

They left as silently as they had come. I stayed for a while, sitting cross-legged on top of the Land Rover. Around the pan all was quiet but behind me, far to the north, I could hear the rumble of thunder and turning around could see lightning flashes. Somewhere between us and Main Camp welcome rain was falling. I clambered down, retrieved my sleeping bag and resumed my interrupted sleep.

We sampled three more stands the next day and the following morning we set off, expecting to arrive back at Main Camp around midday. But by 9 am we were down to our axles in mud. The track we were on had been running along the bottom edge of a wooded slope with just enough clay for a firm surface and the rain I had noticed two nights ago had fallen copiously here, turning it into a Land Rover trap. It was a graphic illustration of a 'catena' effect: the sequence from woodland on deep sands at the top of a slope to a flat grassy area at the bottom, and the mud we were in.

From the results of all our sampling trips we identified 10 main types of vegetation, ranging from teak forest to grassland, each with a typical set of species that varied in amounts from place to place, in all making for a lot of variation. Rainfall was the same everywhere and, although some differences were due to catena effects, most of the diversity was due to three kinds of disturbance: animals (principally elephants), fire and frost (in temperate climes everything is adapted to frost, but not in the tropics). The three kinds of disturbances interact to produce quite different plant communities in the same basic type of vegetation.

In a general way, this is how the patterns of ecosystems develop all over the world. The average climate and the kind of soil determine what's possible – which species can grow – and the species that then actually do grow are selected by the local regime of disturbances: climatic, biotic and fire.

It is the disturbances that create and maintain much of the diversity. Without them, ecosystems simplify: losing plant species and so providing fewer resources for animals. Both the plant and animal communities need exposure to a disturbance to develop the capacity to cope with it. All too often managers try to reduce or prevent disturbances, such as fire, and environmental variation such as availability of water for animals. They create firebreaks and attempt to put fires out, and they put in pumps (such as the one at Mtswiri Pan that wasn't working) to supply water in dry times, and in so doing reduce the patchiness and variety of species and hence the resilience of the ecosystems to the natural regime of disturbances.

In African ecosystems disturbance is very much driven by the big herbivores that sculpture vegetation. Elephants are the biggest and their effects are obvious, converting woodlands into communities of trees and shrubs with a grass layer. But such sculpting effects are not limited to Africa. On some Galapagos islands giant tortoises are the counterparts of elephants, opening up dense shrub and grass areas, creating an ecosystem that favours many birds and reptiles, earning them the title 'gardeners of the Galapagos'.

While striving to reduce natural disturbances, humans have introduced a new kind: interfering with the long-evolved patterns of animal movements. In Hwange, elephants have moved across the Kalahari sands for thousands of years, using ancient paths that follow the lines of Pleistocene dunes. The border between Zimbabwe and Botswana, which cuts across the dunes, was fenced fairly soon after Hwange was established. Botswana and Zimbabwe authorities at the time both thought they had their own elephants, and were constantly accusing each other's elephants of invading their country and causing trouble. The herd that had peeled around me as they moved south of Mtswiri Pan were probably on their way to do just that.

The animal ecologists in Hwange were studying the movements of elephants to establish how large an area they used. Their research budget didn't allow for radio-collars so they devised a system of paint bombs – balloons filled with paint of different colours dropped on animals from a Piper Super Cub aircraft – to identify them in aerial surveys. It was very effective and showed that Hwange elephants moved great distances beyond the Park, both to the east within Zimbabwe and west into Botswana. The fences that had been erected to protect the Park were a hindrance to elephants, as well as to the seasonal movements of buffalo and other animals, and the elephants kept breaking them. For a long time the authorities on both sides were reluctant to accept that they were in fact the same elephants and needed to be managed as a single entity.

The fence between the two countries was ecologically unsound, preventing animal movements in response to seasonal variations and causing abnormally high levels of use, beyond anything the ecosystems had experienced in their evolutionary development. The same thing has been found in other national parks and wildlife regions, and in southern Africa this recognition has evolved into a Transfrontier Parks development, where fences are being removed and parks created across national borders. It has taken a long time to recognise that conserving natural systems requires they be allowed to function across the spatial variation and the range of disturbances under which they evolved.

The extremes of this variation occur in the really long distance migratory species, which require very different habitats in the different seasons of the year. The Arctic tern holds the record with something like an 80 000 km return trip from its northern summer breeding grounds in Greenland to its southern summer feeding grounds in the Weddell Sea south of Australia. Among mammals, humpback whales hold the record with a 5000 km round trip from their breeding and calving grounds in the Pacific to their summer feeding grounds in the Southern Ocean. The migrations of the wildebeests and zebras in the Serengeti, and across Savute to the Kalahari in Botswana, are minor by comparison. But the lifestyles of all migratory species, all of which have evolved over many thousands of years, have conferred on them the benefits of access at different times of the year to environments very favourable to their growth and reproduction. They add to the diversity of species at both ends of their travels but their continued appearance in each is completely dependent on the continuation of the environment at both ends in the forms under which they evolved. If something happens to their Greenland breeding grounds the Arctic tern will disappear from the Weddell Sea.

Disturbance and Goldilocks

The effect elephants were having in Hwange was noticeable, but, apart from areas around some of the water points, it was not of great concern. Mostly it was beneficial by bringing browse within the range of other animals and creating patchiness. In the Sengwa Wildlife Area to the north of Hwange, however, the disturbance was much greater and seemed to be too much, amounting to significant damage and causing much concern. For his MSc, Doug Anderson was investigating claims that the level of elephant damage could not be sustained and he was measuring how much of the vegetation elephants actually ate, as opposed to trashing it in the process. Compared with Hwange, Sengwa is small and fenced off from surrounding areas of subsistence agriculture, and elephant numbers had increased markedly.

One crisp winter morning, Don Parry, our field technician, and I went to help Doug measure the amount of damage on one of his sites. We were walking in single file on a track through rather dense vegetation along the Lutope River. I was immediately behind the game scout who suddenly stopped dead in his tracks and I ploughed into his back, with Doug and Don piling up behind me. We stood peering over each other's shoulders into the eyes of an elephant that was standing facing us on the other side of a tree that had been felled across the track – part of the damage Doug was investigating. This tree had been broken but not killed and was producing new young shoots, and the elephant was feeding on this lush new growth. The leaves on the unbroken trees were less vigorous and less nutritious.

It was difficult to tell who had got the biggest fright. He was so close he could have stretched out his trunk and swatted us. We stood looking at each other for a few frozen seconds and then he whirled around with a high-pitched scream and charged off down the path.

Doug's research showed that, despite the appearance of massive damage accumulating year to year, the amount of damage in any one year was on average only around 4%, which is less than the growth rates of the plants, so on average all should be okay. But the average doesn't mean much in ecology. What Doug also found was that the amount of damage to particular species and places varied a lot. As with many small game reserves in Africa surrounded by lots of people, elephant densities were higher than they had ever been and a few favoured tree species were clearly on their way out.

Were there animal and bird species that needed these doomed tree species? And if they too disappeared how would that influence all the other species? How much damage is okay when it involves selective loss of some species? You can't have elephants without some damage. Is there a point where something needs to be done about elephant numbers – and who should decide that? Who benefits, who loses, and whose values should be used in determining which species deserve more consideration than others?

Much later I was in a meeting in Australia on conserving the alluvial ecosystem of the Murray River, when there was much discussion on where and when to flood a floodplain with water taken back from irrigation allocations. What was it that needed to be conserved? The ecosystem in general? Particular species? Eventually it was decided to target the ecosystem and, in particular, the river red gum – the icon species of the floodplains. Changes in the mix of other species wasn't of great concern as long as the ecosystem functioned well and red gums persisted. The problem was people weren't sure about just what was needed to achieve that, yet the flooding had to begin. In the face of the uncertainty, the way forward was to start the process, monitor what was happening, especially in regard to regeneration and survival of the red gums, learn from that, and adjust the process in an adaptive way. Easy to say, but hard to avoid the bureaucratic urge to lay down fixed rules.

Disturbance of one kind or another plays a crucial role in all ecosystems. It releases bound-up materials, fosters regeneration, increases flows and levels of nutrients and enhances resilience. But in all of them there's a threshold level beyond which the net effects become negative, leading to an amount of change deemed to be too much. And so the same questions that arose in Sengwa apply in all of them. How much disturbance is 'just right'? – as Goldilocks would have asked. (There is, actually, a principle in science called the Goldilocks Principle.) What patterns of disturbance in space and time foster diversity and resilience, and when and how does disturbance lead to their decline?

Diversity in disturbance

I did my PhD research on the prairies of Saskatchewan, trying to work out the ecology of the six million or so glacial potholes, called 'sloughs' (pronounced sloos), dotted across the Prairie Provinces that stretch across Alberta, Saskatchewan and Manitoba. They are small wetlands and ponds left behind as the great ice shield that covered North America during the last glaciation retreated and melted. Vast quantities of sediment settled out, forming the flat prairies of today. So flat and extensive is this region that the saying in Saskatchewan is you can see a man 3 days away if he rides tall in the saddle! In this melting mass as the ice retreated were ice lenses of purer water that melted somewhat later, leaving behind the potholes.

In the first couple of years I noticed that several sloughs had dead patches of a tall grass in the deeper zones. I came up with a hypothesis that the dead patches were caused in years when ice formed right down to the bottom. This, I reckoned, caused freezing down into the mud, and when spring came and the ice started to melt around the edges of the slough, the huge lump of ice floated up, like an ice cube in a glass, tearing the roots out of the mud.

I tested my theory by inserting thermocouples (for measuring temperature) in selected sloughs, at various depths from the water surface down into the mud. They were on long wires attached to a pole I stuck into the mud during the autumn. In winter, as soon as the ice was thick enough to walk on, I started measuring the temperatures, and my hypothesis turned out to be correct. If thick snow fell early in the winter on to recently formed ice, it insulated the pond below and the water didn't freeze all the way to the bottom. This was the normal case. But in one year the snow came too late and the water froze all the way down and next spring the reedy grass in the centre of that slough was dead. The variety of plant communities in the middle of these sloughs was due to variations in the timing of the first big snow falls. A constant pattern of winter snowfalls would result in a simpler set of wetland plant communities.

The sloughs were home to many birds and two species of blackbirds in particular intrigued me while I waded around in the water. I couldn't help noticing them because they dive-bombed me whenever I got too close to their nests. The blackbirds were seasonal migrants who overwintered somewhere down south in the USA. One of them, the red-winged blackbird, arrived several weeks earlier than the yellow-winged blackbird and immediately set about selecting nesting sites in the shrubs surrounding the sloughs. Blackbirds are strongly territorial and defend their nests and surrounding real estate by perching on a high twig at the edge of their territory and singing their heads off. Singing contests accompanied by characteristic head postures determine the outcome of the competition. Trespassers are immediately attacked and the birds ceaselessly dive-bomb transgressors (like me) for as long as they remain in their territories. Being tiny, they didn't pose any mortal threat, but the number of times I got shat on convinced me this was a deliberate part of their territorial strategy.

When the yellow-wings arrived they ousted many of the red-wings from the prime territories around the inside edge of the vegetation, facing the water. Yellow-winged blackbirds are slightly bigger and more aggressive and by the middle of summer there are mostly yellow-wings in the best sites, but with some of the better established red-wings persisting there. Occupation being about nine-tenths of territorial law, the delayed arrival of the yellow-wings gave time for some red-wings to get established in and hang on to sites they would not otherwise have been able to secure. Both the suitability of the sites and the timings of arrivals vary between seasons, and it is this variation that allows the two species to co-exist. If the seasons were constant, one of the species would disappear – it's the 'noise' in the system that is important for them both to thrive, to co-exist.

I found the same thing with two rather similar grass species that grew around the margins of sloughs. They occurred together only around sloughs in which water levels varied a lot from year to year. Where water levels were constant (always enough water flowing in, due to a large catchment) one always out-competed the other.

Back on the other side of the world, in the tropical savannas of northern Australia, a similar thing happens. The climate, and hence vegetation, is driven by the monsoons, and the arrival of the monsoon varies between years. After a long dry season, the much-anticipated first rains can vary by up to a month, from early October to November. Although many species start growing whenever the rains happen to come, a significant number will only grow if the rains come at 'their' time. The set of species evident after an early and late start are a sub-set of all the species. Whenever the rains come there is a set of species able to respond best.

The rare filling of the massive Lake Eyre Basin in central Australia a few times a century are periods of massive reproduction and replenishment for a large array of waterbirds that then spread out to other wetland areas as Lake Eyre dries out again. Without these floods, the waterbird populations would be greatly diminished, or could die out. The same kind of huge floods happens in the Makgadigadi depression in Botswana, which fills with water only on rare occasions. I vividly recall a flight I made from there in a mid-1960s wet period. I took off shortly after dawn and, as the plane climbed up over the main pan, I saw a long line in front, some hundreds of metres from the shore. It turned out to be a line of flamingos several deep and over a kilometre long, feeding in the shallow water. I banked as I flew over them and they took off, turning the surface of the pan pink with the reflection from their outstretched wings in the early morning light: a stunning memory. Had I made the same flight a few years earlier or later it would have been a very different scene: a saline marsh or dry grassland and no flamingos.

The huge variability in climate in these arid regions favours very different sets of plant and animal species at different times, each with different strategies for survival and reproduction. It makes for an ecosystem with high overall species diversity, presenting different 'faces' under different environmental conditions. All of them persist and emerge or re-invade so long as the range and timing of the variability continues. A big concern among conservation ecologists around the world is that under the trend of global warming both of these are changing.

Disturbance and renewal

With its huge diversity in wildlife and its picturesque habitats, the Okavango Delta in northern Botswana is a national treasure. A lot of water flows into it via the Kavango River, 98% of which is transpired or evaporated as it moves through, with a mere 2% escaping out the lower end. All that water makes for lush and productive vegetation in this huge, diverse swamp, but it varies a lot in its different parts, and Fred and Karen Ellery, newly married, did their postgraduate degrees on its ecology – a bit like a working honeymoon. They were investigating what happens to the water as it flows down and how this influences swamp patterns and productivity. Fred studied the channels that coursed through the swamps and Karen studied the swampy parts.

They set up their honeymoon camp on an idyllic island. After their first day of exploration they returned in their boat to find the camp occupied. A large troop of baboons resident on the island had made themselves at home and the camp was a shambles. Fred tore in, yelling and hurling sticks. The baboons fled, shrieking and dropping the bits and pieces of camp gear, clothes and food they had been busily tearing apart. The next day Karen and Fred were careful to put everything inside their tent and close the flaps before heading off. They returned from a day of clambering around on sudds to find the baboons once again occupying their camp. Fred did his yelling and charging act again. The baboons backed off, but less fearfully than on day one. Day three was much the same but this time Fred's charge did not get the desired effect. The baboons regrouped at the edge of the camp and the lead male, a huge individual with very impressive fangs, turned to face Karen and Fred, backed up by a few of his mates. An adrenalin-charged performance by Fred resulted in a second rather casual if not insolent withdrawal. This was no time for valour.

'Let's get the hell out of here' said Fred, turning around to find Karen already packing up. They piled their belongings into the boat and made off, to the silent jeers of the baboons – Fred was sure the big male made an obscene finger gesture as they rounded the edge of the island. They located a smaller baboon-free island nearby and set up camp again.

Fred found that as the water flows along the channels it deposits small amounts of silt, most of it probably arising from land in the headquarters of the Okavango River in Angola (the Rio Cubango as it is called there). Over time, the floor of the channel rises and eventually is slightly higher than the surrounding swamp. Karen showed that in the surrounding swamp decomposition of dead plants gets slower and slower as the supply of new water with nutrients through the sides of the thickening channel margins gets less and less. The vegetation becomes less and less attractive to animals as these back-swamp ecosystems start becoming moribund.

Hippos are frequent users of the channels, moving between deeper permanent swamps and feeding areas around the edges, and every now and then a hippo blunders, pretty much by chance, through the side of a channel, forcing a path into the surrounding swamp. The water immediately starts flowing through the gap, flooding the swamp and re-invigorating growth while the channel beyond the gap is eventually left high and almost dry. The broadly flowing water in the swamp eventually finds a new preferred path and formation of a new channel begins. The pattern that emerges is very dynamic and completely unpredictable.

If the Okavango Delta were to be locked into one set of channels and back swamps, both its productivity and biodiversity would dramatically decline. The swamps as we know them would die. The vibrancy of the delta and its great conservation value are maintained by constant disturbance, and preventing the changes would have disastrous results. Recurring proposals to run a canal through the swamps to get more water out the lower end would decimate the biodiversity of the delta. This is the picture that started to emerge in the renowned Everglades of southern Florida. Efforts there to control the highly variable water flows have had serious unwanted effects. But they have been recognised, and costly engineering aimed at restoring the original flow regime is underway.

A secondary channel made by a hippo blundering through the side of a main channel in the Okavango swamps. These changes in water flows keep the Okavango in a vibrant, productive state (photo: Fred Ellery).

As an aside to this general overview of different kinds of disturbance, what is happening now in the Okavango Basin highlights a very important point: if the disturbance regime that gave rise to the diversity and productivity of an ecosystem is significantly changed it can have dire consequences. Recent diversion of water for irrigated rice from the Rio Cabango in Angola is a major new disturbance that is already having a significant effect on the Okavango Delta. As more water is taken, it will lead to thresholds in required water regimes for various species of both plants and animals being crossed. A number of key focal species are likely to disappear and the current very high biological and economic (tourism) value of the Okavango swamps will be severely diminished.

The effects of a big disturbance mostly tend to fade over time as the ecosystem returns to something like it was before, but sometimes it doesn't and the changed state persists, as it has in a South African savanna for at least a couple of hundred years. The savanna concerned was the focus of an ecosystems project in the Nylsvley nature reserve.

'Nylsvley' is old Afrikaans/Dutch spelling for 'Nile wetland' or 'Nile marsh', and not far from Nylsvley is a spring with the old Afrikaans name 'Den Oog Van Zen Nyl' – The eye of the Nile. It and Nylsvley were so named by the first Afrikaners (the 'voortrekkers') to reach that part of the world. They had been travelling in their ox-wagons for some years by then on the Great Trek, and reckoned they had come across the source of the Nile when they discovered this spring and the magnificent wetland further on.

The basic Nylsvley savanna is sandy and infertile with broad-leaf trees and coarse grass, but within it are several small fertile patches with acacia trees and a layer of nutritious 'buffel' (buffalo) grass. Their origins intrigued us so we invited an archaeologist to come and have a look. At the very first patch he pointed at a piece of rock sticking out above the surface and exclaimed: 'Well, just take a look at that!' To us it looked like a piece of rock sticking out of the ground, so we looked at him expectantly. He looked at us, turned up his hands, and said 'You don't find rocks in deep deposits of sand like this – that is a grinding stone'.

It turned out each patch was underlain by a 'village horizon', down to ~30 cm. This layer was much higher in nutrients and organic matter and, in addition to the odd grinding stone, had pieces of charcoal and old pottery. Below this horizon the soil was the same as in the surrounding savanna.

The charcoal and pottery indicated an age of somewhere in the 14th or 15th century. It is not known when these patches were abandoned, but they have not been occupied in living memory. Each ancient village had an area where livestock had been penned at night. The acacias and buffel grass had grown up on and persisted in these enriched, abandoned patches. But without the cattle providing dung in the pens, soil studies predicted that rainfall should by now have leached the extra nutrients from the dung down and out of this sandy soil, and because the acacias and buffel grass grow poorly in the infertile soil they should by now have been replaced by the normal sandy savanna plants. How had these patches remained for so long?

The answer came from a study on the feeding ecology of cattle and antelopes. Both of them move through the nutrient-poor savanna creaming off the choice bits, but concentrate their time in the fertile patches where they also lie up to chew cud. When they rise the first things they do, in sequence, are arch their backs (a cow's way of stretching), urinate and defecate. The effect is to simulate what happened under the previous human occupants: let the cattle graze freely then bring them in to a few sites where they concentrate the nutrients.

The key point here is that it needed a period of forced enclosure to create the alternate state of this savanna, a long time ago, but, once established – once the amount of nutrients in the soil and the proportion of buffel grass had got above a critical threshold level – the feeding behaviour of animals then maintained the reinforcing mechanism without any further interventions by humans.

Really big sudden disturbances can also have prolonged effects on ecosystems, and someone who had discovered how this can happen was Hugh Raup, who spent many years studying the Harvard Forest near the New England village of Petersham. I got the opportunity to spend a year doing research there and had the good fortune to spend time with Hugh, then Emeritus Professor of Forestry – a grand old character with time on his hands for chatting with a young scientist. He gave me the clue that helped explain some of the patterns we were finding in Africa that didn't fit with the rather linear theories about change at that time. He took me into the forest along a track he had travelled many times, showing me where he had conducted a long-term study with his students.

'To get to grips with the age structure of the forest we cut down all the trees on this hillside' he said, pointing to what was now a reforested slope of young trees. 'We counted the growth rings in all the trees to age them and then examined the forest floor for signs of disturbance.' It was here that he uncovered the key to why the forest looks like it does.

Their examination revealed a large number of humps and hollows. When they measured and mapped them they found there were two groups, one with higher mounds and deeper hollows than the other.

'The thing that gave us the clue to what had happened' said Hugh 'was the orientation of the mounds and hollows. They were all the same within each group but very different between the groups'. The mounds and hollows were where trees had fallen down, the hollow being where the roots had been torn out with the soil enmeshed, and the mounds being the soil. The trees had mostly rotted away. The two patterns of mounds and hollows were a record of two hurricanes that had come from different directions, more than 100 years apart.

'This was an exciting find' said Hugh, 'so we examined other forests around Massachusetts and New Hampshire. We found much the same thing and this allowed us to extend the history of the forests in the whole region back to the 15th century'. Cyclone records showed that any one piece of forest is likely to be blown down, on average, about every 120 years – and this is less than the lifespan of any of the tree species in the forest.

'All of these trees need mineral soil for their seeds to germinate and become established' he explained. 'They can't do that in the layer of leaves and mulch that covers a mature forest floor, and the only time when mineral soil is exposed is after a tree falls down.' The seeds that happen to be near the top of the mounds of exposed soil after the blow down are those most likely to establish and win the subsequent race for light. The plants that largely by chance become established after a hurricane are still there when the next hurricane comes along. All the species, except for those on extreme sites (very wet or dry, very sandy, etc.) are capable of growing well everywhere.

Hugh's research revealed what is now accepted as a general ecological principle: rare, big disturbances are the drivers of real change. They are the events that re-set vegetation and therefore the structure of ecosystems, and the interactions of all the other disturbances then contribute to the variability we see.

Sometimes the vegetation you see in an ecosystem is a transient mix, resulting from not just one but a sequence of different disturbances many years before. This is what happened in Botswana's Savute region, where Petri studied lions and hyenas. It is a magnet for tourism today, but it has not always been so.

In the late 1890s, two significant events coincided: the Savute channel dried up after a long period of flowing, and the rinderpest plague (a virus that kills ungulates) swept down from East Africa, decimating the herbivore populations. Elephants are not affected by the rinderpest but their numbers had been severely reduced by ivory hunting around that same time and, with the drying up of the water, the remaining animals moved away.

The consequence of these two events (three if we include the couple of decades of elephant hunting) was an explosion of acacia seedlings, and in the absence of browsing animals they developed to a size that could survive both fire and browsing as animal numbers began to recover. The period required for the trees to escape browsing (giraffe excepted) is in the region of 10 to 15 years. The stands of trees then continued to grow and thin out and after a few decades they reached a splendid park-like appearance with a grassy understorey that attracted many herbivores, in turn attracting lions, hyenas and wild dogs. It remained in this state for several more decades – long enough for all the people who visit and manage the region to think it had always been like that. During this time the numbers of herbivores were steadily increasing.

In 1952 the channel started flowing again, providing a permanent source of water and herbivore numbers increased markedly, leading to more predators, which led to more tourists. Everyone was happy. However, most acacias are fairly short-lived pioneer trees and by the late 1970s some of them began to die, and the expanding elephant population was also taking its toll. The high numbers of smaller browsers were preventing regeneration of new acacias and there was much concern among Park officials. Elephant culling was proposed by some, closure of water by others. The government wanted the attractive combination of acacia parkland and lots of animals to continue and it was hard to convince officials that that combination was a particular window in time where, in the trajectories of vegetation and herbivores, high numbers of both elephants and large trees happened to coincide. It wasn't possible for this to continue.

Knowing what happens in rare, chance combinations can provide opportunities to make changes not possible at other times. The famed Mitchell grass plateau in northern Australia is an extensive area of treeless grass plains dominated by the productive, high-quality (from a cow's point of view) Mitchell grass. It is the mainstay of the cattle industry in that region. The problem is that, over time, constant preferential grazing on Mitchell grass results in it being replaced by other less nutritious grasses. Once the amount of Mitchell grass gets below some critical level, attempts to recover it by removing all the cattle from a paddock and resting it for a while invariably fail.

Getting Mitchell grass re-established required two rare climate events, one after the other. The first is an El Nino: the change in atmospheric pressure across the Pacific that results in a very dry year or two in eastern Australia. Much of the grass dies, leaving space for new grasses to establish. When the El Nino is followed by a La Nina – the opposite to an El Nino – there is well above average rainfall in eastern Australia and the soil remains wet enough for long enough for new seedlings of Mitchell grass to get their roots down deep enough to get established, and again dominate the other grasses. Knowing about this in advance provides an opportunity for a grazier to remove her cattle as soon as a La Nina is indicated, and so make use of the event.

The richness and diversity of ecosystems are in large measure a result of the pattern of their disturbances: the storms, fires, droughts, frosts and the animals, following occasional major events. If some omnipotent planner managed to prevent disturbances and keep the environment constant (again, all too often a misguided aim of managers), we would lose much of the rich array of species and structural diversity that makes the ecosystems what we love and cherish – and that makes them resilient to all kinds of disturbances. Because the disturbances occur unexpectedly, systems need to be able to cope with them when they do occur. Coping capacity is another term for being resilient and it is maintained and enhanced by continuous exposure to all the different kinds of disturbances across all locations at different timescales under which the systems developed.

Novel disturbances due to humans often exceed the evolved resilience of ecosystems, frequently reaching tipping points into new, mostly unwanted states that can be very difficult, or even impossible, to reverse. In today's changing world, disturbance regimes are changing, and Goldilocks' question becomes much more difficult.

PART III

The nature of resilience in society

6

Coping with life

It's mostly in the mind

'Identity' is a core feature in describing how much a system can change and yet still be the same. It's in the definition of resilience. It is also at the heart of understanding the psychology and behaviour of people, and it's the word that brought me into contact with anthropologist Astier Almedom. She had come across the definition in one of our publications and had contacted me to compare the way we ecologists used it with the way she did in describing the resilience of a person. Astier, who died far too young, was born in Eritrea, became a highly regarded anthropologist living in America and Europe, and did much of her research in her home country.

We met up in Stockholm at a meeting on resilience in society, and she told me about a fascinating study she had undertaken on Eritreans who were living in displacement camps. They displayed a range of abilities in being able to cope with being moved into these camps, and the extremes were interesting. Those who were doing worst and had descended into an almost catatonic state, particularly women, were people who had lived all their lives in the village in which they had been born. In Astier's words 'Having lost all connections to the environment in which they grew up, and the daily routines that had defined their lives, living in this completely different environment they no longer knew who they were; they had lost their identity'. Those who were coping best were nomads: their identity was more complex but whatever it involved it wasn't linked to a particular place, and being moved into a new place had much less of an effect on their sense of identity.

Astier used the term 'sense of coherence' (common in social science and psychology) as being equivalent to 'resilience', which is now widely used in both fields. I rather like sense of coherence because it implies bringing together the things that give your life meaning. And I like the use of 'identity' in describing resilience in both ecology and anthropology. Thinking about how it's used in one field sheds light on its meaning in the other.

With the escalation of stress and mental breakdown in Western society there has been a corresponding rise of interest in psychological resilience. There are umpteen papers and books with multiple definitions and descriptions of what makes for a resilient person. In contrast to the frequent use of 'bouncing back' in the general literature, psychologists actually don't see it like that. They see it as learning, adjusting and moving forward: 'Resilient people do not try to control their lives. They surrender to the flow of the wind. They adjust their sails and ride the next wave of their life'.[12] 'The more obstacles you face and overcome, the more times you falter and get back on track, the more difficulties you struggle with and conquer, the more resiliency you will naturally develop.'[13]

The stress placed on absorbing disturbances and hardships and learning from them is very much in line with the ecological understanding that to be resilient it is necessary to be disturbed, to probe the boundaries of that resilience and learn from the effects by re-organising. In people it involves behavioural responses; in ecosystems it requires responses to the environment and connections between species. In psychology the term 'stress inoculation' is the equivalent of probing the boundaries. Some studies have shown that people with a history of some lifetime adversity reported better mental health and wellbeing outcomes than people with no such history. To quote from one: 'Indeed, drawing from theories of stress inoculation it has been suggested that exposure to adversity in moderation can mobilize previously untapped resources, help engage social support networks, and create a sense of mastery for future adversities.'[14]

Resilience in a person is both a set of traits and a process. You inherit a set of traits that make you inherently more or less resilient: general resourcefulness, strength of character, flexibility and so on (dubbed 'ego-resilience' in psychology jargon). But your actual level of resilience depends on how those basic traits develop; it's a process involving positive adaptation, a capacity that develops over time. If you experience positive emotions it assists in recovering from daily stresses.

Psychiatrists identify a gradient of types of people from what they call 'extrinsic' to 'intrinsic'. We all have elements of both with a balance somewhere in between. 'Extrinsic' people are driven by what others think of them. In today's world, they seek three main things: power, financial success and image/status. They are driven by self-enhancing values and materialism – their image to others – and this is promoted and enhanced by the economic growth philosophy endorsing the 'me-me-me' focus. 'Intrinsic' people have three identifying features: 'self-acceptance', 'affiliation' (family and friends) and 'community feeling' (contributing to the wider world). They have a focus on being inwardly rich.

Psychiatrist Tim Kasser finds that extrinsic (look-at-me, look-at-me!) people have, and report, lower wellbeing than intrinsic people. We are all partly both, and trying to work out where we are along this continuum can be difficult (and revealing). In trying to find his own balance, dealing with the tensions and unhappiness that the extrinsic part involves, the French Algerian author Albert Camus expressed the value of an inward focus most eloquently: 'I realised, through it all, that, in the midst of winter, I finally learned that within me there lay an invincible summer.'[15]

Out of the many lists of things identified as influencing a person's psychological wellbeing, there are a few common ones that emerge as being most characteristic of resilient people. And top of the list is having a sense of humour. Being able to laugh at yourself and your situation, and especially laughing with others, has been shown to reduce stress and to help you cope. According to their biographers, Abraham Lincoln, Mark Twain and Winston Churchill (of the Black Dog fame) all suffered from periods of deep depression and all three used humour as their way of overcoming it. A second very important attribute is having strong connections with others, in two ways: a support system of trusted, positive people who care about you (and with whom you can laugh), and on the other side of the coin you helping and being responsible for others and not just focused on self. A third, important one is optimism: looking for the positive in difficult situations, adopting the glass is half full rather than half empty view of difficult or trying situations. Being flexible and able to embrace and use failure as a way to learn, being able to face fear and leave your comfort zone is another one. And it's related to yet another: mindfulness – paying attention to yourself and your life and asking yourself tough questions.

Living in a society makes resilience in people very complex because it operates at multiple levels. Intervening at one level can have effects at other levels. To enhance resilience in a child, it may be more effective to provide help to a school or the parents than to intervene at the level of the child. And within a level there can be induced secondary psychological effects on other individuals: a psychological disorder in one member of a family can lead to effects in other members, such as sleep disorders. Some of these secondary effects may be suspected and can be dealt with, but often they come as a surprise. In dealing with someone exhibiting signs of low psychological resilience, the automatic focus is on that individual, as a person. But a recurring message coming out of resilience studies in all kinds of systems is that the tendency to focus solely on the scale at which the problem is expressed diverts attention from considering the cross-scale connections that are often the main cause of loss of resilience.

As discussed earlier, ecologists and social scientists use coping capacity as another term for being resilient, but psychologists draw an important distinction between resilience and coping strategies (in contrast to capacity). Depending on how resilient you are, and in what ways you are or are not resilient, you will develop a set of strategies to cope with some particular adversity. Two people with different kinds and levels of resilience will come up with different strategies. And the same person can have different degrees of resilience at different times, and so would adopt different strategies to cope with an adversity. This is an important distinction, because meetings after some disaster, designed to identify what the attributes of resilience are, all too often come up with long lists of personal and community characteristics, most of which reflect strategies that individuals or the community adopted, not general resilience attributes.

A summary of resilience that ecologists have come up with after many different studies includes three main features: (1) absorbing a shock; (2) re-organising and changing in response so as to deal with it while staying essentially the same; and (3) learning from it so as cope better with shocks in the future (the definition in Chapter 1). A summary of resilience that a group of psychologists came up with has three very similar features: recovery, sustainability and growth.

Recovery refers to the extent to which the person regains equilibrium, following upsetting events ... applies to 'a return to normal' following cognitive and affective disturbances in homeostasis that result from stressful experiences. Sustainability refers to the perseverance of desirable actions, goal pursuits, and social engagements that are sources of positive emotion and self-esteem. Here, resilience is measured by the extent to which sources of personal and social meaning in the person's work, family, and leisure life are preserved. Growth, meanwhile, refers to the realization of greater understanding of one's capacities, and new learning that arises as a consequence of the stressful experience and outcomes of one's coping efforts. Although somewhat counterintuitive, many individuals report increases in personal growth in the face of stress; this has been shown to be especially salient in the process of benefit finding.[16]

Virtually all discussions and writings about personal resilience are focused on how to build it, assuming it's always a good thing. But, once again, resilience *per se* is neither good nor bad. Some people develop strategies to build their own resilience by exploiting and reducing the resilience of others. Famous people in history are often held up as exhibiting high resilience, inferring it was 'good'. Some, such as Mahatma Gandhi, Nelson Mandela and Florence Nightingale, deservedly so. For patently evil, highly resilient dictators it is clearly not. But for others it's not so clear. After riding roughshod over people to achieve a desired outcome during difficult times, history tends to focus on the outcome and attribute the leader with high personal 'good' resilience. But, as the saying goes, many famous good people were real bastards in achieving what they did.

But it's also in the body

When resilience in a person is referred to, it is almost invariably associated with psychological effects, but biophysical ones can be just as significant, and they vary a lot between people. As with psychological resilience, we inherit part of our biophysical resilience in our genes and part we develop by the way we live.

We differ genetically in our ability to cope with temperature extremes, but the degree to which we realise that potential depends on how fit and healthy we are and how much we probe our temperature boundaries. If we keep the temperature we're living in within a narrow range and always dress to stay 'just right', we lose some of our innate ability to tolerate, let alone be comfortable within, a wider range. Exposure to high temperatures, for example, has been shown to result in a build-up of heat-shock proteins and to changes in the enzymes that control the rates of biochemical reactions. Exposure to very cold environments leads to an increase in the body's basal metabolic rate (BMR): the sum of all the body's metabolic processes when it's at rest. There is a marked difference in the abilities of people with different BMRs to function under very cold conditions, but the increased BMR declines with lack of exposure to cold.

Our ability to process food is very much a genetic trait. We all fit somewhere along a gradient from being predisposed to easily putting on weight to being thin no matter how much we eat. One classification is from endomorphs (tubby) to ectomorphs (skinny). The evolutionary explanation for the ectomorph type relates to its advantage in being able to run easily over long distances, as in hunting down prey in dry, extensive environments. Tall, ectomorphic people from northern Africa are often the long-distance running champions of the world. Kalahari Bushmen are small and ectomorphic, an adaption to their lifestyle, but the women are different. The ability to store fat, most vividly expressed in Khoisan women, was an evolutionary advantage for them in a very sparse environment with a variable and often limited food supply. Their adaptation is called steatopygia: very large fat deposits in large, protruding buttocks. Drawing on that reserve enables them to produce and sustain babies during periods with little or even no food.

Steatopygia is a nice example that illustrates the relationship between adaptation and resilience. The genetic adaptation occurs in mothers and the effect is to confer resilience on babies and young children, even though male babies never have that adaptation (they don't develop steatopygia). Adaptations are particular features in organisms that confer resilience to particular kinds of stress or disturbance, and they can also increase the resilience of the communities in which they live.

The propensity for fat deposits in female buttocks occurs to a lesser extent in most other human races, as does the ability of males to store fat around their waists. Today's obesity problem is an unwanted secondary effect of this evolutionary genetic adaptation to an environment with limited and fluctuating food supplies. The genetic food-storing ability that promoted resilience in the early development of humans now acts to reduce it in today's Western world of super-abundant food.

With some variation we all inherit a potentially high resistance to diseases. We've been co-evolving with multitudes of them for a long time. But the degree to which our particular potential is expressed is determined by how much our bodies are exposed to diseases at sub-lethal levels. Exposure builds resilience, both to the particular disease in the form of specific antibodies and more generally in terms of a well-developed immune system. It is a highly significant capacity, beginning during birth as a baby passes through its mother's vaginal canal, absorbing bacteria through its skin and taking them in through the mouth, eyes and ears, and continuing this as it contacts faecal matter further down. Its first mouthful from its mother's breast is colostrum, rich in antibodies. The example of compromised immune systems in children prevented from playing in dirt (mentioned in Chapter 2) has been recognised only relatively recently, and it's hard for parents to work out how much and what kinds of exposure are safe and beneficial: how much exposure helps make them resilient without crossing a tipping point where they become ill.

On the other hand, the microorganisms that cause many dangerous human diseases are becoming more and more resilient (in this case, resistant) to the drugs that are meant to kill them. Rising antibiotic resistance is a huge problem. It is happening because the treatment with antibiotics is erratic, or given at too low a concentration, or stops too soon to kill all the bugs. Those that survive have genetically higher threshold levels that must be exceeded to result in death, and they pass on this genetic capacity as they replicate.

Our psychological and biophysical resilience depends on the degree to which our inherited potential is developed, or not developed, by the way we live our lives. And an important part of this involves interactions across scales: the interactions with other people and society, and also the interactions within the body (the organs that make the body function, as a person). A failure or poor performance in one can affect the performance of others and, as a result, the body as a whole (e.g. liver malfunction can cause one or more of high blood pressure, reduced red or white blood cells, and bodily itching, among other problems). Malfunction in the body can cause negative psychological effects in the brain, and vice versa. At scales above the individual, disturbances in society or bad relationships with other people can have significant effects on an individual's mental and, in consequence, bodily wellbeing.

Cross-scale interactions turn out to be one of the most important (and generally most overlooked and least understood) attributes of resilience: in a person, a society, a city or any other kind of dynamic system.

7

Living together in society

Rules for sharing

Every Thursday morning a small group of people meet in a square outside the Cathedral of Valencia. Several onlookers watch as a presiding officer guides a discussion about problems that have arisen around using water. The group consists of *syndics*: representatives of autonomous irrigation communities around Valencia that all depend on sharing water from the Turia River. Meetings like this have been going on for several centuries. The topics discussed and the nature of the problems have changed over time, and the process of the meeting has evolved to suit new conditions. But the persistence of this institution is remarkable and is a major part of why the system of irrigators and irrigation practices has persisted successfully for such a long time.

The Valencia irrigation area is known as a *huerta* and consists of several irrigation communities, and there are four other *huertas* in this region of Spain, each dependent on its own river system. Each *syndic* within a *huerta* uses a communal set of canals built a long time ago and collectively maintained by that community, using their own system of taxes and allocated labour time. The farmers in each community elect their *syndic* to act on their behalf in resolving problems and establishing rules for sharing the water.

In dry times, disputes arise over use of water: some farmers take more than they should, or a farmer may not put in the agreed time for maintenance. In the Valencia *huerta*, a dam was built in the upper reaches of the Turia in 1951 to modify the effects of droughts and floods in this region of very variable rainfall, but water issues still arise. There are temptations to take water out of turn, or too much, but water use is closely monitored. At the Thursday morning meetings the presiding officer oversees the process of resolving the problems and the *syndic* members, excluding the one who is the cause of the problem, discuss the issue and make an immediate decision based on established rules for that canal system. The *syndic* members also meet at less frequent intervals to work out modified or new rules for the current environmental situation or new developments, such as the dam in 1951, or any other problems that might have arisen.

There are lots of details involved in the way the *huerta* functions to keep the irrigation system going through good and bad times, and they are described in Elinor Ostrom's landmark book *Governing the Commons*.[17] She also describes several similar kinds of community rule-based systems for using shared natural resources in other parts of the world, and the quality and extent of her work earned her the Nobel Prize for Economics.

One of the other systems she studied is the *Zanjera* small irrigation communities in the Llocanos area in the Philippines. Spanish priests started recording how this system worked in the early 1600s, but it was operating well before then. They, too, have evolved into a system of independent communities of irrigators who work out their own rules for using water and for maintaining canals, and they choose their own officials – known as *maestros* – to make them work. It's different in many ways from the Spanish system: for instance, farmers own land in both the upper and lower reaches of their river system and some areas are set aside for communal purposes, but it's similar in its broad structure.

Rather different from these two irrigation systems is one of common resource use and sharing that evolved in the 1400s in the village of Torbel in Switzerland. A variable, steeply sloping area with private plots for producing food for the household and fodder for cattle, it has communally owned meadows in the alpine areas for summer grazing, communal forests for felling timber and also use of common 'waste' land. There are rules for sharing and, in the alpine grazing area for example, the number of cows a farmer can send up to the meadows in summer is no more than the number he can feed during winter, and there are substantial fines for breaking such rules. The rights to these communal areas are strongly maintained and newcomers who buy land in the village do not automatically get such rights. From the beginning it was up to those who did have them to decide, collectively, who could become a member.

The examples of shared resource use described by Elinor are systems of evolved rules that have enabled the societies concerned to avoid Garrett Hardin's well-known 'tragedy of the Commons',[18] which highlights the dire consequences of a pasture open to all. In all community-based resource-use systems that have persisted for a long time (in Elinor's words, 'long-enduring common-pool resources') there are very clear and strong rules, and outsiders are denied access and dealt with very firmly if they try. Hardin stresses the importance of having such rules, as shown by the parlous state of many open ocean fisheries.

So what are the rules that matter? By comparing and synthesising the findings from all her studies, Elinor came up with a set of eight generic rules, which she calls principles for managing a commons – though later it became clear that 'principles' was a bit too rigid. The following concise summary of the rules, presented by her in a seminar, is a bit different from how they are presented in her book.

Define clear group boundaries.

Match rules governing use of common goods to local needs and conditions.

Ensure those affected by the rules can participate in modifying the rules.

Make sure the rule-making rights of community members are respected by outside authorities.

Develop a system, carried out by community members, for monitoring members' behaviour.

Use graduated sanctions for rule violators.

Provide accessible, low-cost means for dispute resolution.

Build responsibility for governing the common resource in nested tiers from the lowest level up to the entire interconnected system.

All long-lasting systems of common resource use that Elinor studied include these eight rules in one form or another. The order of the rules reflects the sequence in the development of the system rather than their relative importance – which differs according to the kind of system and the local conditions – and their significance changes with time, especially now with the very fast-changing natural, social and political environments.

At a get-together where she had presented her work, I asked Elinor about this over breakfast one morning. Were the rules still working? In response, she made two significant points. The first had to do with the effects of scale, making the first, second and last rules especially important. In the case of the Valencia *huerta*, with the advent of the European Union there were no constraints on flows of products across borders, and relative differences in prices of produce from outside Spain were having a significant effect on the economic viability of some communities. Their tried and proven system had developed within bounded conditions under a national system of governance and local markets, and it was not very resilient to the effects of changes in the boundary and the new influences of governance at the scale of Europe. The time needed for the *huerta's* rules to adapt to the new scale was much longer than the pace at which the social/political environment was changing.

Some years after Elinor's study, a younger scientist, Sergio Villamayor-Tomás, surveyed *huertas* in the Valencia vicinity and his observations aligned with her findings. He found the *huertas* were not thriving as much as they were in the past due to the ageing of farmers and urbanisation, which meant no generational replacement and young people moving to towns, especially in the *huertas* closest to the towns. Ageing farmers were not as interested in maintaining the *huertas* as they were in the expectation of selling later at much higher prices if their land is urbanised. Furthermore, under globalisation, the Tribunal was becoming more of a tourist attraction and using the philosophy of the Tribunal the *syndics* were preventing many of the conflicts from scaling up to the need for the Tribunal.

The second point Elinor made was that 'graduated sanctions for rule violators' was becoming really important. In her words, there must be 'strong punishment for cheaters'. Its rising importance fits with the rise in the 'me-me-me' society that *laissez-faire*, neoclassical economics has fostered: the shift to a society in which personal gain and greed are acceptable. It is difficult, sometimes not even possible, to use legal means to deal with cheaters, such as cleverly arranged debt procedures that allow such people to prosper at the expense of others. It needs some way of invoking social sanctions that get through the me-me-me barrier – something that the perpetrators can't ignore. Again, Spain offers an example: the Cobrador system.

Set up ~30 years ago to deal with intransigent debtors, the '*El Cobrador del Frac*' (debt collector in top hat and tails) is a system that publicly humiliates the debt dodgers. It involves someone dressed up in a black suit with tails and wearing a black top hat carrying a black brief case with the name *El Cobrador del Frac* printed on it. The cobrador trails the debtor, at a respectful distance, never saying anything or addressing the debtor or responding. If the debtor goes into a restaurant the figure stands outside, looking in, quietly. When the debtor emerges the figure follows him again. In a way the cobrador is acting as the debtor's conscience. He breaks no laws, but everyone knows what he's doing, and the target attracts the kind of attention he/she definitely does not want. The system has a high success rate. I'm not suggesting this as a general model – what works in Spain may not work elsewhere – but some analogous sorts of society-based systems are becoming more and more needed around the world to counteract the rising trend in me-me-me behaviour.

In a complementary way to Elinor's studies, Josh Cinner and colleagues at Australia's Coral Reefs Centre examined long-enduring fisheries in Papua New Guinea, Indonesia and Mexico. They focused on the design principles and came up with three key trends that matched some of Elinor's findings. Despite it being notoriously difficult to define boundaries around marine resources, almost three-quarters of the fisheries did have clearly defined boundaries and also defined membership. Second, all were able to make and change rules: the flexibility and autonomy needed for adaptive management. And third, as with the *huertas*, they generally lacked key interactions with organisations at larger scales, and so were vulnerable to changes at those scales.

Rapid change at a higher scale exceeds the ability of long-evolved social-ecological systems to adapt at the same pace, and therefore exceeds their resilience. This is a common finding. What is happening to the famous Bali water temples and their associated irrigation systems, developed over a very long time, is much the same. The rules for sharing in those rice paddies are maintained by a mixture of religious ceremonies and the reciprocal effects users higher in the system and those lower down can have on each other, as described in an earlier book by David Salt and me.[19] Now, the rapid development of tourism has seen young people leave the villages for towns and the temples and rice paddies become more of a tourist attraction than a necessary food production system.

The shortcomings in effectiveness of the principles described by Elinor and Josh highlight a crucial aspect of developing any such set of rules. They cannot be fixed – and Elinor did in fact suggest them as hypotheses. In trying to come up with a general framework for governance, two of Elinor's close colleagues, Marty Anderies and Marco Janssen, pulled together the work of many others involved in this kind of research and in essence found that the principles are more like best practices than 'laws'. It is great to have them, but they are neither necessary nor sufficient for success. In general, the more successful cases did relate to more of the principles being met, but there are also some where few were met but still did well, and others that met them all but didn't succeed. The principles need to change according to context. The successful systems were resilient in the sense that the time taken to change the rules kept pace with the rate of change in the social and natural environment.

Today that environment is changing very fast: faster than the time needed for systems of shared resources to adapt and cope with the changes. The capacity of social groups to influence the higher scale is very limited and mostly their knowledge of what is happening lags behind the events. Perhaps the increasingly widespread use of social media may increase their rate of response and their bottom-up influence.

Building bridges

Mention the word Camargue (the large delta of the Rhone River) and most people immediately think of white horses. They are one of the oldest breeds of horses in the world and have been used for centuries by 'gardians' (Camargue cowboys) to herd the black Camargue bulls used for bullfighting in this region of southern France. Unlike in Spain, the bull is not killed in Camargue and is actually the centre of attention, rather than the bullfighter. The most famous bulls – fast and smart with lots of spirit – become stars and have long careers in arenas. But breeding bulls and herding them on white horses is not all the Camargue is known and used for.

While the herds of horses live in wet meadows and fringing grassland, in the marshes there is an ancient tradition of reed harvesting for thatching (reed-thatched houses are a renowned feature of the region), and the harvesters have tried to get as much of the delta as possible into a state that favours reed beds. Duck hunting is popular here and the hunters want lots of open water, while professional fishers expect a specific water regime to catch migratory fish such as eels and freshwater fishes. There are areas of irrigated rice in what were once wet meadows, and some wheat in the drier areas. It is also a great region for birdwatching: thousands of migrating birds in the marshes and wetlands, especially pink flamingos, which thrive and nest in the more saline areas, and herons, together making this region on the Mediterranean coast a major tourist attraction. And, of course, there are some vineyards.

The flows and meanders of the water moving through the delta have created a mosaic of fresh, brackish and saline wetlands, marshes and lakes. It has been sculpted over centuries through the creation of dykes, canals, sluices and locks to alter the flows and levels of water to suit the various kinds of preferred uses. Resolving the conflicting preferences has been a difficult, ongoing issue, largely due to the absence of some way to get the different user groups to understand how their demands impact other groups' aspirations. It's a classic collective action issue: a common problem in regional societies all around the world where there are different, preferred ways of using the region's natural resources.

Four of the competing land uses in the Camargue delta of the Rhone: the famous white horses; flamingos (conservation); reeds for the thatching industry; and the black bulls for the bull rings (photos: R. Mathevet).

At the national centre for scientific research in Montpellier, not far from the Camargue, Raphael Mathevet and his colleagues have developed a strong interest in the delta's social-ecological linkages and dynamics and, in particular, the competing demands for the different kinds of water regimes. Their surveys suggested some kinds of users are less influential than others, and persistence of them all in the mix that characterises the Camargue was by no means certain. It became clear that, to avoid intransigent opposition to each other, what was missing was some sort of bridging arrangement between the different user groups, and they came up with a way to achieve this. First, they developed a computer model of the ecology and water regimes of the delta's reedbeds with seasonal changes in river flows, and how the water regimes can be changed by manipulating the dykes, canals, pumps and so on. Then, together with the user groups, they developed what is known as an agent-based model: a model that treats each user as an 'agent' and integrates the ecological and economic models with what each of the agents is doing. Each agent makes decisions on what changes in the water regimes they want in order to further their own interests, and the model progresses in a series of annual steps as the seasons and water regimes change, and the agents adjust their preferences accordingly. Examples include: medium or high water levels to attract ducks in winter for the hunters; emptying freshwater bodies in spring to attract migratory fishes; high water levels in spring for growth of reeds; and low levels in winter for reed harvesting with machines, but no harvesting in heron colony areas; and providing enough water during the breeding season for biodiversity conservation.

This role-playing game reveals the complex set of social-ecological interdependencies of all the user groups. In other words, it shows how a decision by one group influences the welfare of other groups, perhaps causing a significant loss for one or more of those other groups, while resulting in only a marginal gain for the group making the decision. The players have to negotiate with each other as the results of their decisions unfold, and the game serves as a very effective means for connecting the different users to work towards a management policy that keeps them all going. Although the bridging process showed that it does help integrate and satisfy the different users, it needs a continuing effort to keep the process active as conditions and political situations change.

Much farther north of the Camargue, in Sweden, a different kind of bridging arrangement was developed to help resolve conflicts in another wetlands area: an agriculture region around the town of Kristianstad. It is a famous nature conservation area known as the Kristianstads Vattenrike, which means 'water kingdom' in English, noted especially for its cranes. But over the years, by using levees to control flooding, much of it is now also a productive cropping area – and cranes eat grain.

Official government or industry bodies, representing (or seen to be representing) one or other side, were unable to resolve the conflict. Then an independent, small, flexible municipal body was formed, the Ecomuseum Kristianstad Vattenrike, with the specific aim of acting as a bridge between local government, landowners and nature conservation groups. It built a loose, social network of local stewards and key members of municipal and higher level organisations. Through its activities, initiatives such as payments to farmers to compensate for what cranes ate progressed to farmers beginning to value cranes, and organising crane-viewing visits by nature groups. A study by scientists at the University of Stockholm of how this initiative started and developed concluded by saying 'Our results suggest that the EKV approach to adaptive co-management has enhanced the social capacity to respond to unpredictable change and developed a trajectory towards resilience of a desirable social-ecological system'.[20]

The Camargue and Kristianstad Vattenrike have come up with quite different ways to link their regions' user groups, and they are just two examples that serve to highlight the importance of having some kind of viable bridging mechanism in the social part of all-too-often failing social-ecological systems. The first function of such a bridging mechanism is to develop strong, effective communication and interaction between the different users and, once this is achieved, to come up with ways to modify and trade off their different preferences. Only then can successful rules for sharing be developed.

Adaptive living

About 100 kilometres north-east of the Great Zimbabwe Ruins (built in the days of the ancient kingdom of Monomotapa) lies the Bikita District of Zimbabwe. It's a beautiful part of the world with rolling hills and fertile valleys within the miombo woodlands of southern Africa. In its recent history it has been occupied for at least 300 years by the Shona speaking people of Zimbabwe, descendants of the Monomotapa people. They have lived a life of subsistence agriculture, practising shifting cultivation and tending their cattle and goats. Their main crops were originally sorghum and rapoko (a native millet) and then maize was brought in, and later groundnuts and cotton.

I arrived there in 1962 as a Land Development Officer in the then Southern Rhodesian Government. Bikita was divided into three zones, and Zone 1 (my zone) was divided between three traditional chiefs. They ruled through a hierarchy of *sadunas* (head of a group of villages) and *sabukas* (village head). My job was to improve agricultural practices, and getting approval from all three levels for any innovations was essential, and not always easy. I soon realised that my biggest difficulties were invariably where social factors intervened and I was struggling to understand how Bikita society worked. Reactions to my proposals were often puzzling, and left on my own I doubt I'd have got very far, but thankfully I was blessed by having sergeant Rudo as my right-hand man. Then in his forties, he had been in the Rhodesian African Rifles and had had a stint with the British Commonwealth forces in Malaya. He knew the area and he knew how to work the system: a man of the world who, fortunately for me, decided I was okay.

There were two main parts to my job. One was general conservation: mainly preventing soil erosion on arable lands by ensuring the construction of contour banks on new lands before they could be ploughed, and inspecting the upkeep of existing banks. The second was improving crop production and livestock husbandry, and this part was achieved through a group of 12 very competent 'demonstrators', all with agricultural training. They lived among the farmers and spent their time advising and training them. I kept them up to date with agricultural developments and innovations and visited farmers with them to point out various aspects where improvements could be made.

Part of the livestock program was trying to improve the genetic potential of cattle through a breeding selection program, which involved educating farmers about what to look for in a bull and encouraging them to castrate the very obvious surplus. A quick survey revealed a ratio of bulls to cows far in excess of that recommended in my Animal Husbandry II course. There were also large numbers of 'stags': animals with one testicle (products of ineffective castration attempts).

I chose an appropriate area for my first demonstration on selective breeding and arranged to have all the cattle mustered. Rudo and I arrived on the appointed day and I proceeded with my lecture on the finer points of selecting a bull for improved beef production, with Rudo acting as interpreter. It went over rather well, I thought. We moved to the assembled herd where I was to demonstrate the theory and point out unworthy bulls due for castration. Mixing metaphors a little, the first candidate in the 200-odd animals stuck out like the proverbial dog's balls. A skinny-legged, big-horned, runtish bull with distinctive black, brown and orange colouring and an enormous pair of testicles. No doubt about it; he was the one. I pointed him out with suitably disparaging remarks about his conformity and suggested he was a prime candidate for castration. All of this was greeted in complete silence with most of the assembled group looking down at the ground or away from me, and the owner had a stony look on his face. I began to wonder where I had gone wrong and at this point Rudo murmured discreetly that this was a very good time to get the hell out of there. We left rather hurriedly, claiming business elsewhere.

'What on earth was all that about?' I asked Rudo, looking across at him as we drove away. Keeping his eyes fixed on the road ahead, he replied impassively:

'You just told the owner of that bull he should castrate his father.'

'I did *what*?' Rudo then explained to me the business of mudzimu bulls.

When a man dies, a good son will, whenever possible, arrange for a ceremony to have his father's spirit transferred into a young bull. As long as that bull lives, the father will be able to enjoy the life of the bull: in particular the pleasures of mounting as many cows as possible. Clearly this puts constraints on what one can do with such a bull once the ancestor's spirit is in him. One would not, for example, contemplate castrating it later to turn it into an ox: a mere beast of burden condemned thereafter to pull a plough. Consequently, when choosing a mudzimu bull, it is necessary to avoid animals that show future potential as a beast of burden.

Once the ceremony has been conducted, if the farmer later wants to use the bull for something else, the only way to change the situation is to find a bull with exactly the same colour markings and then, after a second appropriate ceremony, the spirit can be transferred into the new animal. It should be obvious, therefore, that a good son would choose a bull with very distinctive markings so that in the event of the son's own early death it would be very difficult for any grandsons, or other family members, to interfere with their grandfather's happiness. The owner of the splendidly coloured bull (which anyone could see was an ideal mudzimu candidate) was very offended by my disparaging remarks and the whole village was incensed by my recommendation he should be castrated. It would be best if we did not go back to that village for a while, said Rudo, so that he could have time to repair the situation.

We drove on in silence for a while, thinking about what had happened, and then Rudo said that, although he could also see there were too many bulls, mudzimu bulls were very important. All the people in the village knew them and respected them. I took that to mean it was more important to respect customs than to get the ratio of bulls to cows in line with my Animal Husbandry II course. He was right, and I needed to be more aware in finding ways to improve herd composition.

The value people placed on cattle and their other resources in Bikita was far wider than just their market value. Some years later I met up with a social scientist who had worked in areas like Bikita and had written a paper on 'The dangers of thinking White'. He showed that the real value of a cow in Shona society was more than four times the price it would fetch at a sale. The cow was also valuable for milk, lobola (bride price), dung (for fertiliser and for making 'cement' floors) and as a draught animal – farmers with few animals use cows as well as oxen for pulling ploughs. In his assessment, oxen and bulls had a similar value but he didn't include a mudzimu value for a bull. Estimating this would be tricky because it was all about social values, not economic.

Much later I became involved in research on ecosystem services, which involves establishing the real value of ecosystems beyond merely producing crops or livestock for the market. Ecosystems filter water and purify it, they buffer floods, recycle nutrients, provide natural pest control, pollinators for crops and various other functions. But when I lived in Bikita, marketable goods were the focus of attention, and unfortunately they are still the only focus for most governments in the world. The Bikita farmers were well aware of the real and much wider value of their animals and their ecosystems.

My position as Land Development Officer frequently got me involved in situations for which I was completely unqualified, often coinciding with a visit to inspect land conservation. On one such visit I was asked to (actually, informed I had to) help arbitrate the negotiations on the lobola for the daughter of one of the farmers. The two families were both keen on this marriage but had been haggling for some time about the lobola and had agreed, without consulting me, that when I next visited they would get me to mediate: a no-win situation. I arrived at the home of the prospective bride's father and, after the usual polite enquiries about our respective health and some comments on the weather, I was told that the land inspection would have to wait because I was needed for a more pressing issue.

I was duly seated on a stool in the shade of a msasa tree with the two fathers on each side of me, and the prospective groom next to his dad. The young bride-to-be was in front, somewhat off-centre as if the set-up wasn't being staged. She was bare from the waist up, pounding maize in a vertical hollowed out log with a thick staff. Her friends were standing around in support, giggling and urging her to greater efforts since this was clearly part of demonstrating her high value. I'd seen women doing this task often and they invariably did it in pairs, standing opposite each other and taking turns to lift and pound in a rhythmic way. But this time it was a solo performance. She too was giggling as she threw everything into the task, breasts bouncing, buttocks thrusting out backwards with each downward lunge, and building up a fine sweat. The groom-to-be was clearly much moved by this performance and couldn't take his eyes off her. He leaned sideways to whisper in his father's ear, no doubt urging him to agree to the asking price. His father, however, put on a good show of looking sceptical and somewhat disinterested while making rather disparaging comments about what I thought was the young lady's feet, which I couldn't understand. My grasp of Shona was still rudimentary and without Rudo next to me, interpreting, I was struggling to keep up with what was going on.

A pot of sorghum doro (beer) was brought to help the discussion. The asking price was three cows.

'Hah!' exclaimed the groom's father, casting his eyes upwards, 'hardly worth one cow'.

I could see I was in trouble, as the two of them turned to me for my opinion.

'Why don't we have another mug of doro?' I suggested, 'while we watch the young lady a little longer'.

Good idea. We filled our mugs and I diverted the subject to the state of crops and the rainfall for a while. The fathers were in agreement on this. There was then a discussion on the relative merits of the two young people, as respectful and dutiful children, and the price haggling then resumed. But I had the sense that both were now committed to a deal and after the third round of doro there was a lot of smiling and jocular remarks. The maiden finished her pounding with an energetic flourish and bottom waggle that left the prospective groom weak and misty eyed, and the lobola was finally fixed (as I rather hazily recall) at two cows and a goat. I departed, feeling distinctly woozy, and nonchalantly accepting their gratitude knowing I had contributed nothing to the process other than by being there. I'd also forgotten about inspecting the lands. Thinking back on that occasion, it was a reflection of the high level of social capital and cooperation in Bikita society at that time.

Doro, I began to realise, was often involved in events that were important in the wellbeing and productivity of Bikita society. I used to love watching doro working parties as I drove by them: the Shona version of working bees. Cultivating a couple of hectares of a maize field all on your own is a depressing, almost overwhelming task, but get 10 of your mates over and a morning of energetic bantering and singing while hoeing in unison and the job's done. The afternoon then involves slaking thirsts with pots of doro prepared for the occasion and a few days later you meet with roughly the same bunch of mates in someone else's field and take up where you left off.

(As an aside, although I've forgotten most of the Shona I learned in Bikita I won't forget doro. It means both beer and a vegetable garden at the edge of a *vlei*

Many hands make light work. An expression of high social capital in Zimbabwe's Bikita District (photo: Brian Walker).

(marshy area). Shona is a tonal language and the two meanings are distinguished by the inflection on the second syllable, which goes up for a garden and down for beer. On one of my land conservation visits I urged a village head to prevent his people from cultivating in the beer itself. Rudo enjoyed explaining my *faux pas*.)

Just as in the Camargue, in communal resource use systems such as Bikita, different groups have different preferences and a high level of social cohesion and cooperation is needed for people to manage them. As social capital increases, the time and costs of getting agreements (the transaction costs) go down. It is easier and quicker to get agreement on using common resources such as water and grazing land. Rules are made and obeyed in societies with high social capital, and trust plays a key role. Without it, agreement can't be reached on changes that are necessary, existing practices become entrenched and it's very difficult to respond quickly when a quick response is what is needed. Trust and good social networks are essential and leadership is a critical ingredient.

Of the three chieftain areas I worked in, Chief Jiri's evinced the best leadership as indicated by higher levels of innovation and cooperation. Disputes in Bikita are resolved by the chief after everyone has had their say – and allowing everyone to have their say is key to successful outcomes. From what I saw of the interactions of the chiefs with their farmers, people were allowed more say in Jiri's area than in the others. Once agreement has been reached on any dispute, enforcing it is an equally important part of leadership and Jiri himself was considered fair. Transgressors of rules and arrangements were firmly and swiftly dealt with by his lieutenant, m'Drum (so named because of his splendid physique – his huge belly resembled a 44 gallon drum), who carried a knobkerrie (wooden club) as his badge of office.

Bikita society did not start out with inherently more noble, altruistic people. It had its fair share of reprobates and crooks, but overall the behaviour I encountered had evolved as a necessary strategy for dealing with disasters and an uncertain future in a very variable environment. The Bikita people personified a resilience approach to life, as opposed to one based on maximising yields. Subsistence farming depended on more than one type of crop and integrating the crops with livestock husbandry. Reliance on just one kind of crop was too risky, and not maintaining sufficient grazing areas would mean there wasn't enough draft power for cultivation, and not enough manure to fertilise the crop lands. Only the combination of a diversity of crops and livestock allowed people to keep producing under varying rainfall and intermittent attacks by various pests.

As the number of people rose, Bikita's evolved system became increasingly squeezed. The area of land under cultivation at any one time, in proportion to the area under fallow, determined the rotation time. If it was too short, fertility declined. The chiefs had to decide when and where more land for crops could be taken out of the grazing areas, which were already showing signs of overgrazing. The system worked up to some critical population level around which there was no margin for error, no capacity to absorb some disturbance such as a drought, fire or disease outbreak. Resilience, which initially had been high, was declining.

During the 1960s, the human population in Bikita had already exceeded the level that could be sustained without the improved farming. Shifting cultivation with natural replenishment of soil fertility during the fallow phase was becoming a thing of the past. Land allocated to cultivation was now permanent, and there was insufficient grazing area to support the number of cattle required to produce enough manure in the cattle pens to restore fertility to the lands. Fertiliser was already necessary to maintain food production.

Governance

While struggling to cope with these increasing strains, Bikita then had to deal with a transformational change in national governance as Rhodesia became Zimbabwe. Authorities changed and the more-or-less autonomous chieftainship areas came under increasing national influence. The new President, Robert Mugabe, was welcomed as a hero and life improved at all levels. To begin with, more resources flowed into the regions and both social (education) and natural (agriculture, dams, infrastructure) resources improved. But it didn't last, because at the national scale the government began to founder and the influence of that spread to the regions. Thirty-two years after leaving Bikita I paid a brief return visit: a depressing experience. There was widespread clearing of trees, no contour banks in the many new, eroding crop lands and the vleis (marshy areas) had virtually gone, with deeply incised channels running through them. They had lost the 'sponge' function, which had supplied a continuous flow of water and buffered the streams and rivers they joined against heavy rains. That buffering capacity had made the catchment system of streams and rivers resilient to such shocks, preventing erosion of the water courses and ensuring a continuous water supply. The rivers were virtually dry – in the middle of the rainy season.

Matching the biophysical deterioration was the obvious decline in social wellbeing. There were no smiling faces, no-one waved as we went past. An inept and increasingly corrupt government had undone the programs of agricultural development and undermined the zestful and cooperative nature of Bikita's society. What had happened is articulated most eloquently by one of Zimbabwe's leading young authors, Petina Gappah. She has translated Orwell's famous *Animal Farm* into Shona because, she says, it depicts perfectly the devastating effect on communities of what's happening at the highest levels. In an interview 2 years ago she said:

> *Zimbabwe was born out of a revolution against an unjust white minority government which oppressed its black citizens, who made up the majority. Black people were doomed never to rise beyond lowly stations. Like that of the animals in Animal Farm, the revolution of Zimbabwe's black majority was a just one. But in the 35 years since independence, the architects of Zimbabwe's revolution, chief among them the country's first leader President Robert Mugabe, have used this very fact to justify perpetrating the kind of abuses they had fought against. Like the pigs in Animal Farm, Zimbabwe's leaders have hijacked a revolution rooted in righteous outrage, not only for personal gain but also to remain in power with no accountability to the suffering people who put them in power.*[21]

The title Petina has used is 'Chimurenga chemhuku', which translates to 'Uprising of animals' (rather than Animal Farm), which will resonate with the many Zimbabweans who are reading her book. In the 2 years since that interview Mugabe has gone and there is a new leadership, and Petina believes the revolution is now trying to correct itself.

It takes a long time for social customs and rules to evolve in an adaptive way to changing social conditions, and governance of the country of Zimbabwe by Zimbabweans occurred overnight.

The Bikita story is one of thousands of how societies work. From many studies, the emerging lesson is that social capital is what determines how well they work, and social capital is primarily determined by three things: trust (shared values), social networks (communications), and leadership. The first two take time to develop; the last is often a vexed problem, especially at the higher scales.

The kind and quality of leadership plays a major role in the progress or decline of a society. There is no better illustration than the comparison of what happened in Zimbabwe under Robert Mugabe to what happened in neighbouring South Africa under Nelson Mandela. The difference underlines the huge importance of the role of power, which can be for good or bad. I like the distinction social scientists draw between 'power over' and 'power for'. To be successful, long-term leadership has to be a process. A retired head of a Philippines government department made this point very strongly at a workshop on climate change a few years ago. In essence, he said that when a society is trying to cope with a big disturbance, the need is for a strong 'follow me' kind of leader – leading from the front. But in times of peace and progress such a leader becomes a problem. What is needed is more of a shepherding from behind kind, keeping the society within bounds but allowing, in fact enabling, novelty and diversity, and so fostering resilience by using her 'power for'. (He didn't actually say the last sentence. I added it as a logical consequence.)

The difficulty with this is, once in power, the 'follow me' kind of leader wants to stay there and resists all attempts to remove him (it's usually a him). This calls for a periodic required change in leadership. Under any kind of government, from autocracy to democracy, for the wellbeing of a society the emphasis is on the kinds and adaptive capacity of its institutions and their governance. And, in the absence of a benign dictator, the best option is a functional democracy. In a discussion about democracy and its failings, using Churchill's famous quote about it being better than all the other options, the philosopher A.C. Grayling stressed that it is 'a continual optimization of options, a noisy process; inefficiency is a great protector of our democracy'.[22]

In Jiri's area, the traditional kind of governance was not really a problem because there were no centralised institutions. M'Drum was the Office of Defence and of Law and Order. In times of threat (encroachment into their lands from a neighbouring Chief's area), Jiri could drum up (pun intended) support from his people to oppose it. Hierarchy was limited to three levels: Chief, saduna and sabuku. The Chief basically knew and was known by most of his people. The people who took over governing the multi-scale state of Zimbabwe, with a complex civil service, had grown up in and inherited Jiri's kind of three-tier society.

As Zimbabwe's transition to majority rule showed, when the rate of change in institutions and governance at the top is much faster than the rate at which lower scales can adapt, especially if the top scale change is transformative, the changes become maladaptive. The wellbeing of the original system loses its resilience, and the transformed maladaptive system can become very resilient.

8

Weathering crises

Learning to recover

There are hundreds of gas pipelines running across Europe, linking up main supply lines with distribution lines. It's a very complex network with lots of nodes where the pipelines interconnect, allowing for redistribution from the various sources of gas. If something (such as a terrorist attack) should cause one of the major nodes to fail in the middle of winter it could have disastrous effects. The European Union worries about things like this and a few years ago I was involved in one of their workshops with experts in civil and military defence and disaster relief, using the notion of resilience as a way of tackling the problem. It was an informative meeting in many ways, but one finding of particular interest to me was the emergence of two worrying thresholds in time, which, together with a few other examples, became known as the 3-day rule. First, the pump station and gas lines need to be fixed within 3 days. If it takes longer than this, it is very much more difficult and time consuming due to increasingly frozen machinery. Second, there is generally a 3-day supply of essential medicines and food, but after that lack of both can have severe consequences.

At a subsequent meeting on disasters in the UK, based on his experiences in Northern Ireland, a military specialist described what happened after an attack by the Irish Republican Army. If order could be restored within 3 days, recovery back to how things were before the attack occurred fairly smoothly. If the disorder lasted more than 3 days, new groupings had formed and consolidated, with new connections between them, and between them and the general population, and this was very hard to undo.

In the rural city of Horsham in Australia, an assessment of its capacity to deal with catastrophes revealed a 3-day supply of treated, potable water. Beyond that, all drinking water would have to be boiled. At a workshop in Melbourne to examine how resilient the city would be in the face of another heatwave (the one in 2014 resulted in double the average number of weekly deaths and more than 500 extra emergency call-outs), a 3-day rule in regard to food supplies and medicines at refuge centres emerged.

Threshold effects are generally associated with the amount of something (availability of labour, number of disease cases before an epidemic ensues) or relative amounts (debt:income ratio for a business to succeed), but thresholds in time can be just as important. In any assessment of resilience in the face of a disaster, the question about thresholds in time needs to be asked, and where one is suspected in everyday human systems the 3-day rule is probably a good starting point. However, in the loss of mate recognition in David Westcott's North Queensland golden bowerbird populations (Chapter 3), it is at least a few years. Time thresholds are some function of the kind and the size of a system and, although they are clearly important to know about, it seems they are seldom considered until after they have been crossed.

Hurricane Sandy struck New York in October 2012 causing 43 deaths, damage worth more than US$19 billion, and 2 million people to be without power for several weeks. It served as a tipping point, leading to the explicit inclusion of increasing climate change risks in the rebuilding effort. The city has now integrated a dynamic approach into its climate action strategy in which each response action is evaluated along a flexible pathway of adaptation. Any fixed program of response, especially if it linked all parts of the city in a whole-of-city response, could well lead to a whole-of-city failure. A key lesson was that flexible adaptation strategies need to be locally appropriate yet regionally coordinated, not a patchwork but 'a regional fabric of resiliency'. Keeping critical functions, such as power, water supply and transport, modular increases the resilience of the city as a whole.

A fascinating and more complex example of a cascade of failures (although Italians didn't find it fascinating) was the huge power failure that hit much of Italy on 28 September 2003.[23] A network of dozens of power stations across Italy are interconnected and each one is connected to several internet nodes to which they supply power. Each internet node in turn connects back to several different power stations, and the computers in the internet nodes control the operations of the power stations. On that day a failure and shutdown in one power station led to the shutdown of several internet nodes, which then led to a shutdown in several more power stations that were dependent on them, leading to several more internet nodes collapsing and so on, leading to massive power failure. Cascading failure is a consequence of one part of a fully interconnected system failing. Again, modularity in complex, interconnected systems is an important component of being resilient.

In highly interconnected systems (of whatever kind), some connections are more important than others and, for those that are critical, having a diversity of ways to keep the connection is an important resilience attribute. In 1998 there was a huge explosion in Melbourne's natural gas plant, cutting supplies for 20 days. Quite apart from the inconvenience to Melburnians, in Shepparton (160 km away), the main hub of Australia's milk industry, millions of litres of milk had to be poured onto farmland because a single gas pipeline from Melbourne was the only source of power for the three milk processing factories. They now have other kinds of back-up power generators – demonstrating response diversity.

The frequency and severity of all kinds of crises – self-induced (inappropriate design and engineering), climatic (storms, droughts, floods, heatwaves), disease (epidemics and pandemics), financial, societal (e.g. terrorism) – is on the rise around the world, and social scientists are busy analysing them to help societies cope with them. The word 'cope' is interesting because, although it is widely used to describe the capacity to deal with a crisis, it is just the first of four components that social scientists define as constituting social resilience.[24]

Coping capacity is seen as a reactive capacity, a measure of how a society deals with the immediate impacts of a crisis, and it is a function of several attributes: the quality of existing plans for action, available human and other resources, and quality of leadership, among others. The second component of social resilience, adaptive capacity, is proactive, its main feature being learning. It is the ability and the preparedness to deliberately and explicitly learn from each event, and to have this embedded as part of the society's response to a crisis. It's an ongoing process, incorporating knowledge from big and small incidents and from crises elsewhere.

The third component of social resilience is transformative capacity. This depends on having or creating new ways of doing things, of making a living, and it also entails accessing help and assets from higher levels. Together they lead to fundamental changes in the way the society is structured and works in regard to crises. The fourth component, very much at the heart of social science, isn't so much a particular attribute but rather a mix of human agency, social networks, power relations, institutions and discourses. Collectively they determine how well the other three can operate.

A really important development in the social science approach to crises and resilience, matching so well with the ecological approach, has been the recognition and emphasis of uncertainty, change and crisis as normal, rather than exceptional. The world is perceived of as being in a permanent state of flux.

As a final observation, even though crises occur at all scales in societies at all stages of development, it appears that as a society develops and grows it becomes less adaptive and more and more susceptible to shocks. What's more, history tells us this is not new. It is seemingly an inevitable outcome of how we define progress.

Progress to collapse

Some years ago I accompanied Joe Tainter, an expert on castles, on an exploration of a very old castle on the shores of Lake Balaton in Hungary. We had taken a day off from a meeting of the 'Balaton Group', which grew out of the Club of Rome (which produced *The Limits to Growth*) and still meets at Lake Balaton, in Hungary. Its members started to meet there when East could not meet West and Hungary was one of the few places where scientists from both sides could get together.

Joe's expertise is old civilisations and on the day of our visit he wanted to see a particular ruined medieval castle built on top of the ruins of a Roman fort at the edge of the old Roman Empire. He wrote a great book, *Collapse of Complex Societies*,[25] in which he describes a fascinating pattern of changes and disasters in past civilisations.

As we clambered around the castle, Joe explained the conclusion reached from his findings (and I hope he forgives my very brief interpretation of all his work). As societies grow, they are constantly being confronted with all kinds of environmental and social problems. In attempting to solve them a society becomes more complex. It makes more rules, more conditions, imposes more kinds of taxes, creates more layers of government, establishes more committees, more categories of things to deal with, and so forth. And complexity costs. Eventually the next added costs exceed the benefits they are meant to produce and the system breaks down.

For a society to be resilient it needs, in Joe's words, 'reserve problem-solving capacity' and this gets whittled away by increasing complexity. He described how the Roman Empire, like many ancient civilisations, began to break down at the edges where the increasing demands for inputs to Rome exceeded the perceived benefits of being part of the empire. The old fort under the castle we were in had been at the edge of the Roman Empire. Joe and other historians have chronicled the collapse of more than a dozen ancient civilisations, on all continents, and they show similar, seemingly inevitable, pathways.

The tendency for societies to better themselves and to establish rules to reduce uncertainty and avoid disasters, at all levels from neighbourhoods to nations, inevitably leads to increasing rigidity and a declining capacity to cope with uncertainty and disaster.

Out of the failures and successes in dealing with shocks and disturbances across a variety of social and built systems, a few important attributes of resilience emerge. The earlier piece on how social scientists view coping with crises covered aspects of this, but an overview of societal resilience is worth emphasising.

First and foremost is the mindset and capacity to think ahead, to be aware of looming changes and being prepared to deal with the unexpected. The opposite of this (once ascribed to a former Australian prime minister) is the notion of driving forward by looking in the rear-view mirror. While the forward-looking mindset needs to be shared across the whole of society, it is strongly influenced (as the Australian PM showed) by the kind of leadership. I once participated in a management course at Oxford and was treated to a presentation by Sir Peter Cadbury, then retired. He described how, at one point in its history, Cadbury's was looking down the barrel of failure, prevented by several factors from making the necessary changes they had identified (local and government regulations, union limitations, and others). 'We needed a crisis' he said. 'So we created one. We joined with Schweppes.' The immediate period of chaos following the merger provided a window of opportunity through which the necessary changes could be pushed, and the new, transformed company flourished. The key to being able to take advantage of any crisis is to be prepared, to have a plan of action to put into effect when the crisis/opportunity occurs. It doesn't take long for the crucial period to pass, and then it's too late.

The power, or agency, to act is a second important attribute, and it is strongly influenced by the quality and trust in social networks. A third is endeavouring to develop support systems (power, water, food supply, transport, medical) that are structured and that function in such a way that they are able to recover from a disaster – in other words (and this is a hallmark of being resilient), that will be safe when they fail, rather than being built as supposedly fail-safe systems. And, finally, something that flows from practising the first one (thinking ahead), is avoiding the temptation to choose some particular desired future and to adopt instead an adaptive approach, learning to move forward along a pathway that avoids the kinds of society that are not wanted while keeping open the options for good ones. There are others, but these four serve as a concise account of what constitutes a resilient society.

PART IV
Nature, society and resilience

9
Unintended outcomes

The rule set Elinor identified for long-enduring systems for sharing resources (Chapter 7) required knowing what it was that people wanted and then coming up with the necessary rules to achieve it. Devising the rules was based on their knowledge, their understanding, of how the rules would translate into what they wanted. Although this seems obvious, without a long learning process the knowledge that is used is often a real problem. Sometimes it's because the focus is on only the direct effects of a rule or a management action, without considering possible secondary effects, and sometimes it's because the existing understanding just isn't good enough.

Fiddling with ecosystems

In the world of wildlife conservation, a common problem facing managers revolves around the decision whether or not to cull animals, on the grounds it's either good for them (the wellbeing of their population) or for the ecosystem they are in. It's a tough call: managers get damned if they do and damned if they don't. If they cull and it turns out it wasn't necessary, they get criticised. If they don't cull and other species are lost or there is soil erosion, they are taken to task for that. The grass–wildebeest–lion story in Kruger National Park in South Africa in the 1960s and '70s shows how the dilemma can unfold.

Being selective feeders, wildebeests in Africa thrive under dry conditions when grass is short and nutritious. They form large herds and their impact on the grass is very noticeable. In wet periods the grass is high, there is plenty of water around and they break into smaller groups, so are more vulnerable to predators. Also, food quality goes down as the grass gets coarser and the wildebeests lose condition, becoming more susceptible to increased parasite loads that build up in wet periods.

Wildebeest numbers in Kruger had been building up from low levels around the turn of the century following the end of the rinderpest outbreak all over southern Africa, and also the cessation of hunting in the area. Predator control had been a policy until ~1960 to allow game numbers to increase. By the mid-1960s, the number of wildebeests in the central region of Kruger had grown to ~14 000. It was a period of low rainfall and the main wildebeest concentration areas were looking bare and degraded. Park authorities were unsure whether the high numbers were 'natural' or a consequence of increased water supplies. Should they do something, or wait and see how things unfolded? But then visitors started to complain about how bad the area looked. This clinched it. Culling was implemented in 1965 and continued until 1972, by which time numbers had fallen to around 8000.

In 1971 a period of high rainfall set in and the dynamics of the wildebeests changed. The 1972 cull started, but the numbers were low so culling was stopped, but, much to everyone's surprise, the numbers then continued to decline year by year, down to less than 5000 by 1978. With hindsight, it is possible to piece together what happened. The wetter conditions from 1971, coupled with the low number of wildebeests, resulted in mostly tall grass. Recruitment of new wildebeests decreased because of poorer food quality and because there were more lions, which now had the added advantage of the wildebeests being dispersed in smaller groups. When there had been just a few large herds, territorially dominant pride males kept other lions away.

The result was that in 1971 the combination of reduced birth rate and increased predation meant that more wildebeests were dying than were being born: population growth crossed a threshold from positive to negative so the population continued to decline even without the culling. Park authorities now tried to increase short grass areas by burning the tall grass and stopping buffalo culls (buffalo herds create short grass areas through their grazing behaviour). The effects were not strongly evident. At the same time, following the end of predator control in 1960, lions had increased markedly and by 1975 there were ~700. Was this too many lions?

The ecologist who was studying the situation, Butch Smuts, called for more time before any action was taken. 'There are lag effects in these dynamics and we shouldn't interfere any further right now' he stressed. But, despite his opposition, the authorities decided that in the interests of the wildebeests they had better cull the lions, and 335 of them were shot between 1975 and 1978. A few years later an outbreak of mange (a disease of the skin) occurred in lions and concern about the smaller number of lions, many now in poor condition, was so high that it led to lions being darted and bathed to control mange.

It's easy with hindsight to criticise the Park authorities, but, at the time, faced with incomplete information and strong public pressure to 'do something', it's not surprising they took the actions they did. It was a classic case of cautious fiddling: the inevitable outcome in situations bedevilled by inadequate knowledge and an approach that only considers the direct effects of immediate concern and doesn't take into account secondary changes. That kind of management no longer happens in Kruger Park. Some 30 years later at a symposium there on ecosystem management, I was most impressed by the approach, the understanding and philosophy that had evolved. What had happened in the intervening years in the development of ecological science and management to induce that change (not only in Kruger but in many places around the world) is a big part of learning where, when and how it is necessary to intervene, or not intervene: how to build resilience.

Providing animals in a game reserve with pumped water during a drought to keep them alive and available for viewing (and Kruger has been careful about this) has often had the unintended secondary effect of severe overgrazing and soil erosion, and the loss of animals that need tall grass. There are also plenty of examples from using chemicals to control pests in natural and agricultural systems, best illustrated by the rise and fall of DDT. And there's a fascinating example of this that ends up with parachuting cats into Borneo forests.

In the 1960s, malaria was a serious problem in Borneo and the World Health Organization set up a DDT spraying program. The immediate effects were good: malaria significantly decreased as mosquito numbers fell. But the DDT also killed other insects that were eaten by lizards and skinks, which in turn were eaten by cats (there are many native cats in Borneo as well as domestic) which then died, leading to a surge in rat populations.

The rats became a big problem, eating rice and stored food and leading to a surge in sylvatic plague and typhus among the Dayak people. So domestic cats were rounded up in towns and villages and transported into the forests. In some remote inaccessible regions, this involved parachuting bags of cats into the forest from low-flying airplanes.

Pests and guests

These stories show how unexpected things can happen when, due to inadequate understanding of delayed and secondary effects, you do something to get what everybody wants and it ends up with something nobody wants. The world of ecology is replete with such unintended consequences, very often arising from introducing new species into an ecosystem – and in terms of really bad outcomes Hawai'i probably has the worst record for this.

The Hawaiian Islands Land Trust is trying to figure out if a resilience approach to their problem might help restore their remaining native species and, on a visit there in 2016, the Director, Scott Fisher, showed me around the island of Maui. At one point he stopped and said: 'Look around you. There's not a single native plant species out there. Every tree, shrub, grass and herb has been introduced from some other country.' It was really surprising to me because, from a landscape perspective, it looked like a diverse, healthy ecosystem. And many of the animal species are also foreign. It is no longer possible to restore the original ecosystems of Hawai'i. They have been transformed to such an extent that the change is irreversible. So how does resilience apply here?

The Trust is trying to conserve some remaining icon species, such as waterbirds that are hanging on in some places. They were decimated by rats that arrived as stowaways with the first Hawaiian people, and in an attempt to get rid of the rats mongoose from India were introduced. The problem is that rats are nocturnal and mongoose are diurnal so seldom the twain did meet – but the mongoose wreaked havoc with bird nests.

In Hawai'i today, resilience 'of what' must identify new, transformed ecosystems that Hawaiians like. Some of them, such as particular wetlands, still have desired native bird species in them and the task is to work out how to make them self-sustaining populations as integral parts of new resilient, self-organising ecosystems.

Hawai'i is worst off in terms of introduced plants and, although it also has big problems with rats and weasels and other pest animals, in terms of unwanted outcomes from introduced animals Australia takes the cake. It was the thing that struck me most forcibly when I arrived to join the CSIRO Division of Wildlife and Ecology. A big part of the Division's research was on how to control these pest species, originally introduced with good intentions but now competing with, and all too often eliminating, Australian species. Never having encountered anything like them before, the Australian species had little ability to combat or evade them, and thanks to these invaders Australia has the worst record in the world for extinction of native animals.

Hindsight, again, makes it easy to understand. Australia was a very strange place to its early British settlers and to make them feel more at home they brought with them many of their favourite plants and animals. The Acclimatisation Society was formed in the very early days with the aim of making this strange new land look and feel as much like the mother country as possible. Over the years the Society brought in ~6000 different species. And other species, thought to be potentially useful in this strange place, were brought in from other countries. In time, various domestic livestock escaped and became feral so that today Australia has very large populations of feral rabbits, hares, foxes, cats, pigs, donkeys, camels, brumbies (horses) and various other small mammals, birds, reptiles and fish. The term 'brumby' harks back to one Private James Brumby who came to Australia with the New South Wales Corps and later settled on a free land grant where he bred horses. In 1804 he joined an expedition to Tasmania and turned his stock loose. He may have started the process, but many more horses were released into the wild with the decline in the Indian army remount trade after the First World War, and today there are more than 300 000 brumbies in Australia, the largest population of wild horses in the world – North America by comparison has 40 000.

Brumby control is a major problem, and it's controversial. My son Sean spent a year as a jackaroo (the Australian version of a cowboy) in the Coopers Creek region of Queensland and one thing that upset him, as a lover of horses, was the number of fine-looking brumbies that got rounded up with the large mobs of cattle and sent to the abattoirs for pet food. It's all part of Australia's history of the mixed successes of introduced plants and animals. We have more camels than Saudi Arabia. European starlings and many other foreign bird species are displacing native birds. Carp from Asia now threaten many Australian native fish species. Cane toads were brought in by the Queensland Government to control a grub that ate the roots of sugarcane. The toads ignored the grubs, ate large amounts of native insects and spread all over the tropical parts of the country. They have poisonous glands on the back of their necks causing major losses in predators such as the northern quolls. Though the net effect of cane toads on Australian native fauna is distinctly negative, they have now become part of Australian culture. The cult movie *Cane Toads* remains popular, and dried cane toad skins are made into purses for sale to tourists.

On the other hand virtually all of Australia's agricultural wealth depends on animals and plants brought in from other countries. For the most part they have been well behaved and so are welcome, but sometimes the situation is not that clear and people are divided in their views. At a meeting of the Australian Conservation Council a discussion on whether or not to eliminate introduced trout from Australian rivers raised the point that we need to distinguish between pests and invited guests: trout were considered to be in the latter category. This idea applies quite widely. In the earlier discussion about the ecosystems inside us (Chapter 1), the good bacteria that keep our digestive systems working are guests and the tapeworms, blood flukes and so on are the pests.

In the long list of Australian pests, the archenemy has always been the rabbit. On Christmas Day in 1859, twenty-four of them arrived from England, imported by Thomas Austin, a member of the Acclimatisation Society. Thomas put his rabbits in enclosures on his property 'Barwon Park' near Geelong in Victoria and, as rabbits do, they multiplied. Some were soon released (or escaped) and within a decade they had become a pest over large areas of Victoria. In 1866, a mere 7 years after the first ones arrived, 14 253 rabbits were shot on Barwon Park. They spread rapidly and by the turn of the century had crossed the Great Divide and reached the eastern seaboard. Forty years after their introduction, they occupied most of the continent south of the tropics.

Australia has put an enormous effort into trying to get rid of rabbits, by all sorts of means. In 1887 the Colonial Government of New South Wales offered £25 000 in an international competition (worth about $10 million today) for an effective rabbit control method. Louis Pasteur tried to claim it through a demonstration that chicken cholera would kill rabbits, but it didn't work in the field and he didn't get the reward. A man by the name of Rodier advocated and attempted to implement the 'Rodier' method: to whit, trapping lots of rabbits, killing the females and releasing all the males. The theory was that the sex-starved males would harass the remaining females to death. It didn't work. Another suggestion that predation by cats would solve the problem actually led to a train load of moggies being transported and released into the outback where they wreaked havoc among the remaining local wildlife, adding to the feral cat problem. The £25 000 prize was never awarded.

Rabbit control gave rise to the 'rabbit-ohs': a whole new occupation and associated workforce dedicated to shooting and poisoning rabbits, and ripping and fumigating warrens. It has occupied millions of man hours and dollars with success being limited, at best, to a holding operation. A biological control agent was needed and in 1950 the myxoma virus was imported.

Myxomatosis is a disease specific to rabbits, causing a none-too-pleasant death. There was public concern about whether the virus really was specific to rabbits and in the early stages of release an outbreak of encephalitis in humans in the Murray Valley sparked fears that it was caused by the myxoma virus. In an exemplary display of confidence and public spirit, the chief executive of the CSIRO, Ian Clunies Ross, the eminent scientist Sir Macfarlane Burnet, and the virologist who had worked on the disease, Dr Frank Fenner, publicly injected themselves with the virus. Nothing happened to them and public fears abated.

The virus eventually took off and by 1954 it covered most of the southern two-thirds of Australia, killing millions of rabbits and leading to a resurgence in vegetation and the wool industry. But the spectacular effect did not last because mutations in the virus gave rise to less virulent strains that rapidly increased. The highly virulent strains killed their hosts quickly, leaving less time for mosquitoes and fleas to sample the host's blood and so pass the virus on to other rabbits. Also, rabbits began to develop immunity to the virus as the few that were able to survive passed on that ability to their offspring. The net effect was that rabbit populations began to increase and by the early 1980s plagues were again occurring.

Then a new virus that killed rabbits, the calici virus, appeared in China. It was imported into Australia after ensuring that it could not reproduce in anything other than a rabbit, and it again spectacularly reduced rabbit numbers, but not everywhere. And now it, too, is losing effectiveness and rabbits are once again on the rise. Nevertheless the cumulative value to Australia of these two viruses in terms of avoided losses due to rabbits is more than $70 billion, so research into new means of combating rabbits will most surely continue. But the lesson so far is that rabbits in Australia are very resilient to whatever 'disturbances' are thrown at them.

The one part of Australia rabbits can't invade and survive is the northern hot tropical region. But Asian buffalos can, and they arrived in the 'Top End' in the mid-1800s and quickly became feral. As is their wont, they (literally) wallowed in the wonderful array of wetlands and billabongs, many of them covered with beautiful red lilies, and converted them into messy, muddy wastelands. In the 1970s and '80s there was a campaign to get rid of them. As carriers of bovine brucellosis and tuberculosis, they were a threat to the cattle industry, they were clearly having a bad effect on native biodiversity, and in any case they made the landscape look awful. So ranchers, conservationists, tourists and tour operators alike demanded a removal program. 'Get rid of the buffalos!' was the cry. In response, CSIRO conducted a trial to examine the effects of buffalo removal at a place called Kapalga, next to Kakadu National Park. The area was fenced in half, the buffalos removed from one half, and what happened in both was then compared.

One interesting result was that, after the buffaloes had been removed, the wetlands did not just return to how they were before. They were invaded by tall, ugly weeds that would have blocked the view for tourists in the Park who were sitting in a vehicle. It looked worse than before. 'Bring back the buffalos!' Fortunately good sense prevailed and after a few years ecological succession gradually replaced the weeds with the species characteristic of these wetlands. 'Red Lily Pond' hadn't seen a red lily for decades but just 10 years after the buffalos were removed they were back. The lesson for impatient managers is that the recovery pathway of an ecosystem is generally quite different from its pattern of decline. After a big disturbance you should not expect an ecosystem to come back straight away to how it was before. It may well come back along a different and longer path.

The Northern Territory was declared free of bovine diseases in 1997 and culling buffalo ceased outside national parks. Their numbers resurged and they are now valued guests because they are an important part of tourism, including hunting safaris, especially in remote areas such as Arnhem Land. Values change. So there are now new rules for how buffalos should be managed. More generally it's hard to predict how the trade-offs between the welfare of native biodiversity and the commercial value of introduced species will play out in the future.

Water weeds are introduced pests in lakes and rivers around the world, and in the tropics and subtropics a really bad one is a free-floating aquatic fern from South America, *Salvinia molesta*. Some years ago Australia's tropical lakes were covered in dense mats of this uninvited weed that had somehow arrived from South America – where it didn't cover lakes. Mechanical and chemical controls failed but then a weevil, *Cyrtobagus salviniae*, that had evolved with it in South America, keeping it at low levels there, was brought in. It reproduced rapidly on the

Salvinia, decimating it. Its own numbers then crashed, but as the *Salvinia* started to recover it followed suit, reducing the amount of *Salvinia* again, and so the two oscillate around a low level of each. In Papua New Guinea, however, where the weed had also invaded lakes, introducing the weevils failed: they didn't reproduce and died out. An Australian scientist, Peter Room, studied what was happening and recommended liberal fertilisation with nitrogen. The reaction from authorities was one of amazement. 'What?' they asked. 'Just look at it – it's growing like crazy. Why would you want it to grow even more?'

Peter had found that the level of protein in the *Salvinia* leaves was just below what was needed for *Cyrtobagus* to reproduce. The waters of the lakes had very low nitrogen levels and his experiments showed that with added nitrogen the protein went above that threshold level, the weevils reproduced and expanded and the *Salvinia* was brought under control. What then surprised everyone, including Peter, was that there was no need for any further nitrogen fertilisation because, unlike the mature leaves, actively growing leaves had enough protein for weevils to reproduce. The one-off fertilisation was enough to get the 'system' across the threshold into the low-weed/low-weevil state, and it then self-organised to stay that way. This kind of effect is not uncommon. It's the same kind of process that keeps the fertile acacia patches going in the sandy Nylsvley savanna in South Africa

Before and after. A lake in North Queensland that was covered by an invasive aquatic weed *Salvinia molesta* from South America (left side), and after the introduction of a weevil from South America that feeds only on it (photos: P. Room).

(Chapter 5). Once the patches had crossed the threshold for keeping cattle and wildlife attracted to them, the animals no longer needed to be penned at night. The fertility difference is now maintained by the animals' changed feeding behaviour.

Harvesting the tropics

When looking at a map of Australian vegetation, the amount of tropical rainforest seems very small. And it is if you compare it with the Amazon or central Africa, but at the scale of north-east Queensland it is extensive. Logging the forests and associated timber processing was a big industry in the region, but it ended in the mid-1980s after intense clashes between foresters and conservationists. In one of these, the Minister for the Environment went up to the north Queensland region of remaining forests to explain the policy, and in a heated exchange copped a punch from an irate forester. Being a generally unpopular character, this provided a moment of accord between the opposing sides, but eventually logging ended. It has had positive results and much of the biodiversity in these forests will be conserved. In the longer term, however, as presaged by what is happening in the golden bowerbird populations studied by David Westcott (Chapter 3), there may well be further losses and tackling them remains on the list of conservation problems.

The logging in Australia has ended, but for most of the world's tropical forests it is still very much the go. In some, it's just a case of selling trees for timber but mostly it's about clearing the forest for agriculture, which raises two big questions: can tropical forest regions with their very high rainfalls be farmed sustainably, and are there any likely secondary consequences from clearing them?

In the late 1980s, the first of these questions was examined in a forest just outside the small town of Yurimaguas, in the upper reaches of the Peruvian part of the Amazon Basin. Pedro Sanchez from the University of North Carolina conducted an experiment to find out what happens to soil nutrients when rainforest is cut down for agriculture. It was part of an international program and Pedro organised a meeting at the research station there, which was quite an occasion for Yurimaguas, with a couple of dozen international scientists descending on the town. All eyes were on the station and, recognising their moment, the local workers went on strike, lodging a list of demands. On the opening day the director arrived at the locked gates, where we scientists were waiting, and began trying to reason with the workers who stood impassively with folded arms, confident they held the best hand.

Sweet reasoning gave way to veiled threats and the chief of police was called in: an impressive man, very large, not given to long conversations, a big gun on his hip. He stood by and watched the negotiations for a while then slowly took a couple of steps towards the strikers. The strike ended and the workers unlocked the gates, smiling at him rather nervously. Still without saying a word, the chief of police climbed back into the passenger seat of his jeep and was driven away. The scientists proceeded to the research station and the meeting began.

We met the chief of police again that evening at a reception given by the mayor, who had assembled the town council in the town hall. He was seated on the stage next to the mayor, arms folded across his ample chest with his usual thoughtful look. The mayor made a speech in Spanish, translated by a young graduate student. He began with an impassioned tirade gesticulating vigorously and pointing outside, then at us, and every now and then up towards either the ceiling or heaven. He paused, nodding to the graduate student, and we all turned to her with interest to discover what the mayor felt so strongly about. She stood up with a big smile and said: 'He say "hallo".'

Looking a bit disconcerted at the brevity of the translation, the mayor doubled his efforts. The student kept the ratio of translation to presentation time about the same and the mayor eventually lost interest in her and concentrated on the Spanish-speaking members of the audience. At the end of his speech he sat down, the chief of police leaned forward, and the local members of the audience burst into enthusiastic applause until he sat back again.

In the reception that followed we learned that the mayor did indeed feel strongly that something had to be done to find a way of increasing food production. The area had to satisfy the demands of the long-established population as well as an influx of people from over the mountains, which we took to be Lima and its environs (it was the Andes, not heaven, he'd been pointing at). He was depending on Pedro to provide some answers.

Current wisdom was that trying to farm the Amazon intensively was a costly failure, resulting in rapid loss of all the nutrients leading to leached, degraded soils that couldn't sustain agriculture. And what's more, they couldn't be returned to productive soils: the Amazon ecosystem was not resilient to agriculture, and the degraded state was very resilient to efforts to reverse it.

The Yurimaguas experiments showed that this was indeed true if the forest was cut down and the soil left bare for any length of time. However, if the soil was never left without a cover of actively growing plants no loss of nutrients occurred. Timing was all-important. Without active roots, the high rainfall resulted in massive leaching of the nutrients that were constantly being released from decomposing organic matter. Pedro had developed a system of continuous cropping in which the new crop was planted under the old one, and at the time of our visit had produced 22 continuous crops of rice–rice–beans–rice and so on in 7 years. There was always a cover of actively growing plants. We examined soil pits in the cropped area and adjacent rainforest and compared their profiles and the results of soil chemical analyses. There had been no loss of nutrients.

So, farming the Amazon isn't necessarily a disaster but it requires sophisticated and careful management to avoid degradation, and, given the prevailing management and relaxed attitudes to life, a less demanding and less risky approach was also investigated. It involved cutting down and burning the rainforest but not totally clearing the land. The stumps of the trees were left and a less intensive practice of cover cropping and under-planting without ploughing was used – again ensuring there was always a cover of actively growing plants. The yields were less but nutrient loss was still prevented. The high-tech system was very vulnerable to any hiccups in management: it had low resilience. The low-tech system was less productive but more resilient. For the high-tech approach, a chief of police, or perhaps a m'Drum, stationed on every farm would do the trick, but such upholders of attention to detail and enthusiasm for the job are rare.

The results from Yurimaguas showed how you can increase the production from an ecosystem without losing its resilience, provided management is up to scratch. And, taken on its own, this was rightfully regarded as a big benefit because it reduces the need to keep cutting down more forest. But it can't be taken on its own because increasing production has a secondary effect. As crop yields go up they reach a threshold level above which commercial agriculture becomes economically viable, and so it spreads. Using tropical rainforests for agriculture therefore doesn't only have effects at the scale of the farm. Pedro's quest for a sustainable way of farming – one which did not involve continually clearing more forest – in fact can make clearing forest more profitable and so has important consequences at higher scales. This brings us back to the second big question about using tropical rainforests: the secondary, indirect consequences of clearing them.

Around 80% of the rain that falls in the Amazon is actually water recycled back up into the atmosphere by the trees, through transpiration. The original Amazon forest was established when the world was much wetter, so it didn't need the recycling to get going. As rainfalls decreased to today's levels, however, the recycling/pumping action of the forest has become critical. It is the recycling that has been able to keep overall rainfall high enough, in fact still more than enough, to maintain the forest. But there is a threshold level of forest below which the amount of recycled rain will be insufficient to ensure that total rainfall is enough to maintain a forest. Coupled with bigger and more frequent fires, the forest will then change into an open woodland and eventually a savanna, with comparatively little transpiration, and then further secondary consequences will start because a significant amount of the water currently transpired by the forest moves south of the forest area (due to air circulation patterns over the equator).

These aerial flows of water have been dubbed the 'Amazon's flying rivers' and are estimated to carry as much water as the Amazon itself. As the forest gets cleared, that source of rain will be hugely diminished and the subcontinent will become drier. The recycled high rainfall is a service the rainforest currently provides, and beyond some critical level of deforestation the whole region will be on a trajectory of declining rainfall, unable to recover.

A comparable scaling-up of local interventions in a forest was the unexpected outcome of spraying the spruce forests in Canada to control the budworms that Buzz Holling studied (Chapter 3). Under natural conditions the forest had developed such that there was a dynamic mosaic of outbreak patches and patches in various stages of recovery. The outbreaks were irregular and patchiness had been induced a long time before by differences in weather effects and fire. Under such conditions, the forest persisted as a spatially dynamic mosaic of patches in different stages of development and decline.

With all good intentions, the foresters intervened with insecticide to control the budworms wherever they were approaching outbreak levels. For a while it seemed to be working, with more of the forest in a well-developed state. But as the number of areas sprayed increased, so the total area of forest now in the same mature state increased. The picture wasn't all that clear because of the variation in climate and fires and spread of budworms, but it seemed the foresters were now stuck with continuous spraying: if it were to stop, the whole forest would die from the resulting massive outbreak of budworms. After a lot of proposals and panels of experts, the solution was a combination of logging and sequentially stopping spraying in parts of the forest to get the pattern of stages in growth, collapse and re-establishment back to the natural proportions.

Coming back to the tropics, although the fate of nutrients is the key issue in high rainfall forests, in the dry tropics the critical effects are all about water, and in particular the effects of livestock grazing on what happens to the rain that falls on rangelands. It's a two-way thing: the amount of rain that gets into the soil determines how much grass can grow, and the amount of grass cover determines how much of the rainfall goes in. Rob Kelly studied how this works by comparing several dry savanna areas used in different ways in south-eastern Zimbabwe, and he showed that the big differences were actually due to the amount of litter that covered the soil. Under litter there are lots and lots of small pores in the soil, made by all kinds of insects, especially termites. Ants prey on the termites and also make holes into their nests below the surface. All this activity in the interface of the soil surface and litter makes for a mini-ecosystem of its own, meshing with the soil ecosystem below. Remove the litter and all that disappears.

Rob showed that rain went into the soil 10 times faster when covered by grass and litter than it did on bare soil. On bare soil, rain drops 'bombed' the fine soil particles loose and they settled out to form a sealed surface, like a clay tennis court, with clogged up pores, and the rain ran off instead of into the soil. Such soil 'scalds', as they are known in Australia, can last for decades. The litter ecosystem determines how much water penetrates and, in explaining what was happening while we were crouched over one of his sample plots, examining the litter, Rob described something that had at first puzzled him.

'I found that infiltration on bare areas was actually lower where there were no animals compared with where there were some. But when I took a closer look it was clear that having a few cattle and goats or impalas walking over the bare surface was good, because their hooves chip the sealed surface and allow some rain to penetrate. And I also then found that where there is litter', and here he pulled up some half-buried pieces, 'animal hooves help to push it onto and into the soil. But if you have too many animals you end up with bare soil. It's a tricky thing to get just right.' (Goldilocks, again.)

It turns out there is a particular, threshold amount of bare surface that is critical. If grass and litter cover is above it, the amount of rain going in, rather than running off, allows increased growth, producing more litter, and so more of the rain goes in and, as long as grazing is then less than the amount of growth, the ecosystem restores itself. Below the threshold, things get progressively worse.

In really arid regions such as the Sahel, crossing this threshold leads to desertification. Once it is crossed, even if there are no animals, the desertified state is very resilient and it takes a prolonged wet period to get the rangeland back above the threshold.

Long distance effects

In addition to the evidence of major forest blow downs, another thing Hugh Raup showed me in the Harvard Forest were several old stone walls that run through it. They are evidence that the whole region was once an agricultural landscape of crops and pastures. All the forest on land that could be ploughed had been cleared and the stones removed. In the 1830s and '40s, what is now the Harvard Forest was owned by John Sanderson, a prosperous farmer. The region was at that time dependent on an agricultural economy and what happened to it is described in a wonderful paper by Hugh, 'The view from John Sanderson's farm'.

In 1830 the Erie Canal was opened, providing cheap and quick transport for the rapidly expanding supply of crop products from the mid-west farming country – much more productive than the lands in Massachusetts and cheaper to grow. Within 20 years, agriculture on John Sanderson's farm was uneconomical and abandoned. The first trees that grew back were the early successional white pine and they were then used for making barrels. Hugh grew up in Ohio and remembers, as a boy, being sent to buy salt mackerel (which he loathed) from such barrels in the local stores. The barrel business lasted about as long as it took for the forest to achieve its mixed composition of oaks, maples and other hardwoods, and the land-based economy of the Petersham region declined to almost nothing – a secondary consequence of all the benefits of the Erie Canal.

Writ large, the story of the Harvard Forest is the story of re-forestation in significant parts of Europe, now economically non-viable for agriculture as a consequence of cheap products from cleared forests in the Amazon and other third-world countries. Europe still pays large subsidies to its farmers to stay on the land but much of the food in the supermarkets comes from lands cleared for agriculture in the tropics: dictated by simple economics (with the emphasis on simple). The uncertain fate of the Valencia *huertas* (Chapter 7) is a Europe-scale long distance effect: a seemingly inevitable outcome of globalisation effects on markets and associated urban drift.

A somewhat different version of this kind of secondary effect is currently a worrying trend in South-East Asia. Because cutting down tropical forests has bad ecological consequences, an international movement arose, banning importation of timber that was not certified. However, the secondary effect that wasn't foreseen was that, instead of just harvesting trees for timber, whole forests are now being replaced by oil palm, which *can* be imported – no bans on it. Economists call this the rebound effect, and in this case it is leading to a transformation from a harvested forest to agriculture.

These long distance socio-economic effects bring to mind the vulnerability of migratory species to changes at one or other end of their annual journey (Chapter 5). Their long-evolved successful lifestyles have virtually no resilience to such a change.

Our genetic and social heritage

The unintended consequences of efficient (cheaper) agricultural production go beyond the negative economic effects they have in other parts of the world. They include the loss of many traditional varieties of crops and livestock that are now out-performed by a few breeds and varieties demanded by consumers in other countries. Fifty years ago very few people thought about the world's genetic heritage and even fewer had any concerns about it. Today the losses in it are all too apparent and getting worse. There is a wide-scale decline in the genetic diversity that confers resilience to changes in the environment, and to pests and diseases. A classic early example was the 1840s Irish potato famine: a genetically uniform crop exposed to a new strain of blight disease and a secondary effect of the famine was the mass migration of Irish people to America. A more recent example was the loss of a big fraction of the Asian rice crop to the grassy stunt virus in the 1990s.

India originally had an estimated 400 000 varieties of rice. These had declined to 30 000 by the mid-1900s and, following the Green Revolution, thousands more have disappeared. The International Rice Research Institute in the Philippines is aware of the dangers in this and has a program for planting out and maintaining local varieties of rice – but it's a huge task. The same goes for wheat and corn and in fact for all the big food item crops.

The problem is largely due to the fact that the global seed industry is monopolised by three transnational corporations. In their own interests, they concentrate seed into a few varieties most profitable to them and their shareholders, oppose practices such as saving and replanting seed, and thereby diminish the future resilience of global crop production. China has over 50 unique pig breeds but many are now endangered as Western breeds replace them. Many of the high-performing selected varieties could not survive on their own. All commercial turkeys in the USA now have to be artificially inseminated because the breasts of the males are so large they cannot copulate.

The great uncertainty around the inevitable, looming changes in agricultural environments (climatic, diseases, economic) makes this decline in our genetic heritage a deeply worrying trend. If left to continue, a really serious unexpected outcome of this trend is likely, and it'll be too late then to lament 'we should have known better'.

The social counterpart of what's happening to genetic diversity is the decline in social diversity, captured by the fate of languages around the world. In Australia there were over 400 languages. Many have gone and some are now only spoken by a few people, or even the last remaining speaker. Of the officially recorded 853 languages in Papua New Guinea, the most linguistically diverse country in the world, 12 are extinct and it is estimated 124 are in trouble and 40 are dying. One can ask, does this really matter? People will be able to communicate more widely and understand better what is happening in their country and the world. That's good isn't it? But it comes at an adaptive cost.

Indian author Virek Shanbhag grew up speaking Kannada, a 1000-year-old language (at least). He is fluent in English but writes in Kannada because there are no meanings in English for lots of words and phrases in Kannada: the insights and implications of a word or phrase are literally lost in translation. So he uses a translator and together they have translated his novel *Ghachar Ghochar* as best they can. 'I write in Kannada' he said during an interview at a writers festival 'because the next word just comes to me. There is a fragrance to a language'. That fragrance embodies the evolved adaptive value of the language, enabling the society to interpret and understand the social and biophysical context of, and so contribute to its resilience in, the complex environment where that society lives and where its language developed. As the world faces an oncoming stream of shocks and novel kinds of social-environmental combinations, could the collective set of languages act as an adaptive resource?

Climate change

In the face of mounting denialism in politicians and corporate leaders – so blind they will not see – it's appropriate to end this account of unintended outcomes with a brief comment on climate change. To do so, I'll briefly describe just one that is now beginning to happen. Climate change itself is of course an unintended outcome of the growth in human population, the global economy and the consequent increase in consumption and processing of natural resources. Most people have heard of the huge changes it will cause in weather events, in agriculture, in water availability and in the movements of diseases, and most now understand and accept this. But new secondary effects are arising. One is the consequence of increased photosynthesis due to increased carbon dioxide. The concentration of carbon dioxide in the atmosphere has increased from 280 ppm in the middle of the last century to over 400 ppm, and it is rising to a projected 550 ppm in 50 years. The increase is already causing increased plant growth. This is used by some opposed to the short-term costs of doing anything about climate change as evidence that it's a 'good thing' (as stated by one US senator).

However, it has been shown in experiments that the levels of carbon dioxide in the atmosphere expected by the middle of this century will cause a significant increase in carbohydrate content in most grain crop plants, with a resulting decrease in the concentration of nutrients, especially iron, zinc, potassium and phosphorus, and also in protein. Iron and zinc are already limiting in these crops, with about two billion people suffering deficiencies, causing a massive decrease in life years (earlier deaths), and the projected increase in deficiencies will result in a huge global health problem.

The costs of doing something about climate change can be calculated quite accurately, and they are immediate, so they dominate short-term responses. Though the costs of not doing anything will be much greater, they are not easy to quantify and, because the really significant effects will occur some time in the future, they are all too easily discounted.

10

Growing pains

Actually, it's not 'the economy, stupid'; it's 'the stupid economy'.

Development

In 1860 my great-grandfather John Walker left Botany Bay in Sydney to find his fortune. He made his way north working as a fencer and then spent several years as a tree clearer before marrying my great-grandmother and becoming the publican in the village of Wellingrove. He was part of the massive Australian effort to clear the 'scrub' and open up the land for agriculture. He fathered 11 children as part of his effort to develop Australia and was no doubt as proud of his tree-felling as he was of his large family. Life must have been fairly hard though, because his fifth child Leslie left home while still a teenager to join an Australian contingent going to South Africa to fight in the Boer War. By the time the boat arrived in Port Elizabeth in 1902 the war had ended, but my grandfather stayed on in Africa.

John Walker would have been astonished to learn that more than 100 years later his great-grandson would be involved in a program to re-establish the woodlands he had sweated so hard to clear. He and his compatriots had no idea that their tree-felling efforts would lead to soil salinisation problems. The delayed secondary consequences of tree clearing were ever thus. From Gilgamesh's day to the present, people have over-used natural resources. Why? What drives them to go beyond what's good for them? It conjures up the words of the theme song from the movie *Alfie*, used as the opening for the book.[26]

So is it just for the moment we live? When you sort it out, are we meant to take more than we give? It seems we still don't understand that we can't keep taking more than Earth can replenish; we have to take less than what Earth, with our help (our giving back) can regenerate. Starting from 1 January, Britain consumes all of the food it produces in a year by Easter. It cannot survive on its own resources. At a global scale, the Global Footprint Network calculates that by 8 August 2016 the world had used all of what it can produce and regenerate in 1 year. For the rest of the year we were operating in overshoot by depleting stocks of fish, trees, soils and other resources and accumulating waste.

Why do we continue doing what is clearly unsustainable? Four reasons for not changing our behaviour stand out:

- ignorance, in spite of the fact that the information is available
- pure greed and short-termism (to hell with others and the future)
- being trapped in a way of living from which it is very hard to escape
- the absence of any consequences to those who are doing it – there are no negative feedbacks to their actions.

A greater effort in public education and awareness will help overcome ignorance, and there are encouraging signs of this across multiple media. But, again, there are none so blind as those who will not see, and this unwillingness to acknowledge the evidence in front of them is mostly due to the second reason.

Greed is a growing outcome of the culture promoted by economic rationalism that places individuals above society on the disproven belief that this is best for the economy, and therefore everyone. Countering it requires refocusing on the wellbeing of society rather than advancing 'me' as the goal. Social justice researchers talk about 'othering': to make themselves feel okay about their undeserved, elevated position, 'me' people distinguish between themselves and 'the others'. There is no guilt in knowing you are better off than the others. Yet even if it is 'me' you're most concerned about, societal wellbeing is the major determinant of your wellbeing.

The third reason has to do with being stuck in a way of doing things that is really hard to change. Escaping such social traps involves short-term costs that cannot be borne by just a few, and so they can only be escaped if everyone, or at least some critical mass, starts doing it together. The difficulty lies in how to initiate the transition to something that is going to be radically different. The classic example today is how to change from the pursuit of short-term economic growth as the overriding measure of progress; the kind of progress that is assumed necessary for an increase in quality of life for all. Effecting the set of changes that may achieve at least maintaining quality of life requires a different set of progress indicators, and achieving this will require a change in our culture.

The fourth reason is lack of feedbacks. If a factory dumps waste into a river without any consequences to the factory owners, all the consequences are borne by the river and those using it downstream. There is no feedback to the way the factory operates so there's no reason to stop dumping. Though financial penalties for dumping would solve the problem, all too often governments would prefer companies to maximise profits and so contribute to increased national gross domestic product (GDP). But apart from regulations, social feedbacks can be effective and, from local up to national levels, this is gaining ground.

Internationally, regulatory feedbacks are missing or largely ineffective, often obscured either by a time lag between cause and effect (as in the cases of Gilgamesh and John Walker) or by distance and evidence. Until recently, shoppers in European supermarkets buying meat from chickens fed from grain feed that includes soy bean grown in the Amazon had no idea their purchases were causing more rainforest to be cleared for more soy beans. I say until recently, because in this particular case NGOs concerned about what was happening in the Amazon – the icon of NGOs – got into the act. Through a media campaign using remote sensing images, they showed the world the alarming rate of rainforest clearing and put pressure on the global corporations doing it. It was an effective feedback and since 2006 the major transnational corporations involved in soy trade signed a moratorium on purchases of soy being produced on farms causing deforestation in the Brazilian Amazon.

But there was a secondary effect of this positive outcome. The feedback to the producers doing the clearing was in connection to their Brazilian Amazon forest activities, so soy production shifted south to 'deforestation havens' in the Cerrado region in Brazil and the Chaco biome in Argentina and Paraguay and thus far there has been little NGO clamour about that. Few people know or care about it

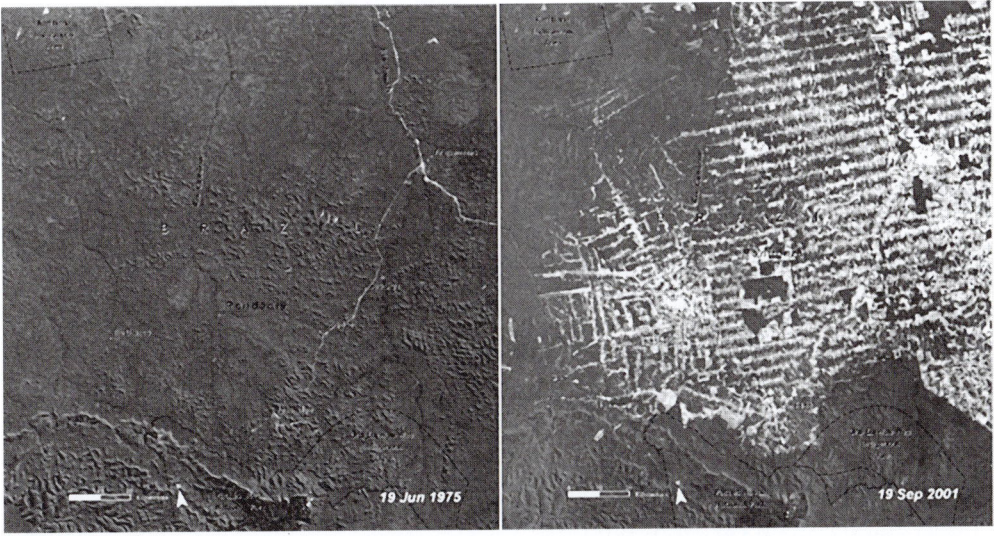

Remote sensing of forest clearing in the Amazon Basin over the period 1975 to 2001. Images like this were used in the NGO media campaign to put pressure on the corporations responsible for the clearing. (photo: UNEP, n.d., Rondônia. Environmental Change Hotspots. Division of Early Warning and Assessment, United Nations Environment Programme).

and the data showing what's happening have not been picked up by the international NGOs and media. The consequences may affect regions that are less iconic than the Amazon, but they are still very significant.

Multinational corporations work largely outside national laws, and with weak or unenforceable international regulations most of them experience no negative feedbacks from the way they operate. With few exceptions (such as the Amazon campaign, which attacked the reputation of big branded companies) they are little constrained by values and ethics as they pursue their objective of maximising quarterly profits for shareholders. A few have seen the light, to protect their reputation or pre-empt public regulations, and in their own long-term interests are trying to change, but most of their many competitors are not.

As already described for climate change at the end of the last chapter, secondary consequences that occur out in the future are really problematic because their costs are hard to specify. One can readily put numbers on the immediate costs of taking precautionary actions now (reduced profits, job losses, etc.), but it's very difficult to estimate the long-term (and therefore discounted) costs of not taking such actions now. So the short-term prevails.

The four reasons for not changing our behaviour have combined to put us on a trajectory to inevitable declining human wellbeing, globally. We can't keep on growing the human population and at the same time use more resources to increase, or even just maintain, everyone's standard of living. Population levels in developed Western countries are levelling off, but the world population is not, and per capita consumption of natural resources is going up everywhere. Over 200 years ago Thomas Malthus explained why this can't go on, and in 1972 *The Limits to Growth*[27] spelled it out quite clearly. Right-wing ridicule of the book at the time has now itself been ridiculed by an analysis of what actually happened in the subsequent 35 years, showing we are indeed on the 'most likely' path the book predicted, yet the aim of most governments is to continue trying to do it. The question is: can we stop it, ourselves, before it gets stopped by one or other catastrophic versions of the looming nexus (introduced in Chapter 2)? Now appearing on some corporate agendas, the nexus is between declining availability of water, energy and food. They are tightly interconnected and a crisis in one will lead to crises in the others, triggering human disasters, disease epidemics, migrations, wars and revolutions – and a collapsing global economy.

Hopefully, as awareness grows and a shift in culture strengthens, the world will be able to take sufficient action in time to avoid the nexus. It will likely take a rare, big event/crisis/opportunity to trigger this; though our history in seizing such opportunities isn't too encouraging.

On 10 November 1989, together with thousands of other amazed and excited people, I was pushed and pulled up on top of the Berlin Wall – the day the Wall 'came down'. I was in Berlin at a meeting to finalise the International Geosphere-Biosphere Programme on climate change. A week earlier, arriving on a Sunday with an afternoon to spare, I had taken a conducted bus tour into East Berlin: a grey, unsmiling experience.

Now, just 5 days later, the communist East European system had collapsed. It had begun late on the Thursday night. Inexplicably, the incredibly tight border controls had suddenly been relaxed. Our meeting had gone on into the night and as we walked back to our hotel sometime around midnight we noticed Trabants (the not very good East German car) driving along Kufurstendamm, West Berlin's main drag, and were puzzled by this. We woke the next morning to discover that Berlin had been transformed. We closed the meeting early so we could witness this momentous occasion and headed for the Wall.

Hordes of people were streaming towards the Brandenburg Gate from all directions in West Berlin. At the base of the Wall, a young man with a hammer was smashing at it with all his might while grateful spectators (me included) picked up fragments as souvenirs. The top of the Wall was packed with people of all ages giving vent to the most emotional experience I have ever witnessed. To my right a young woman was digging at it with a Swiss army knife and further along a percussion band had people doing a sort of constricted samba. Close by on my left an old man stood with his arms outstretched towards the Brandenburg Gate in the East German side, with the Wall making an arc in front of it. Tears were running down his face. On the East German side stood a line of soldiers. Some were tentatively smiling up at the crowds and they were clearly uncomfortable.

The space between the Wall and the Gate was empty except for the line of soldiers but beyond the Gate was a crush of East Berliners. Despite the exuberance in the night just a few hours earlier, they now seemed somehow subdued. People were still coming through at Checkpoint Charlie further along the Wall but the line of soldiers held the crowd back from surging forward to join their West Berlin brethren atop the wide part of the hated Wall. Turning towards West Berlin, all I could see were streams – rivers – of people flowing towards the Wall, surging up to it and mingling with the crowds that were clambering up and down like a highly active ant colony.

Why had the breaching happened on that night, in that week? It was of course not coincidental. It had been brewing for a long time. The socialist system of government and business was not working. The disparity between East and West was growing all the time and it was plain for all to see that, whatever the original intent and theory might have been, the reality of a communist socialist state had had its day and had to go. It had very little resilience and needed only a small shock to push it into collapse. The cracks in the system had appeared well before that

November. A crucial step had been the decision by Hungary to open its border with Austria, thus creating a conduit out of East Germany. The swell of discontent, the accumulation of objections and malfunctions reached some critical threshold at the end of that summer as discussions in the East tried to find a way to open the border. Books have been written about it, but perhaps no-one will ever know how it was that on that particular night the sudden change occurred. There was confusion between different checkpoints and somehow those at Checkpoint Charlie were allowed through. Whatever the actual trigger, the change was inevitable. Its time had come and the tipping point would have been triggered by something else, very soon, if it hadn't happened that night.

Revolutions such as this only take place when the weight of evidence in favour of the new paradigm becomes overwhelming – often when the dominant protagonist of the old paradigm dies (to paraphrase Max Planck: 'Progress is made funeral by funeral'). In the overthrow of communism, the arch-protagonists had long since died so it wasn't an event precipitated by death of a leader. The revolution took place when the balance of evidence against the communist paradigm and in favour of democracy and modern capitalism – not a new but an

The day the Wall 'came down'. Planning Panel of the International Geosphere-Biosphere Programme against the Berlin Wall on 10 November 1989 (author third from left). Was this upheaval a lost opportunity? Could it have been a turning point for the start of a sustainable global economic system? (Photo: Brian Walker)

alternative paradigm (which as we have seen also has problems and now needs its own revolution) – could no longer be ignored. It was the communist paradigm that was dying, and it was then that ordinary people gained nerve they had previously lacked and those in power lost their own convictions about the infallibility of the communist system and their ability to keep the proletariat in line.

On top of the Wall that cold and windy Friday afternoon in November 1989, the shift in that balance was palpable. The third reason for not changing (lock-in/social traps) had passed the critical mass, the tipping point needed for change. What followed the collapse of communism determined the new trajectory of the world. It was a rare, big opportunity. So how did the world re-organise after that event? What trajectory did it adopt, and what other trajectories could it have adopted?

Twelve years later, on 11 September 2001, crowded around a TV screen with dozens of other horrified passengers in the departure lounge at Stockholm airport, I watched a replay of the second plane as it flew into the World Trade Center's second tower. In the 12 years following the fall of the Berlin Wall, the world had experienced a period of peace, free of terrible threats to global security the Cold War era had bred. Now this had happened. What had transpired during these 12 years?

The West could have used the period after the Berlin Wall collapsed as an appropriate time to reflect on what was needed to ensure a self-sustaining, long-lasting system of global governance, resource use and economics. Indeed, many thought leaders did just that, but political and corporation leaders instead consolidated the West's hold over the global economy, pushing forward the era of free markets, globalisation, economic rationalism, privatisation of public assets and deregulation. Neoliberalism became the dominant trajectory.

Economic growth and global trade were given priority because they would lead to a better life for all. The so-called 'trickle-down effect' would somehow make everyone better off, justifying the huge profits and salaries at the top end, and the market would sort things out. But real economists have repeatedly shown the trickle-down effect to be a myth: favouring the rich does not produce a rising tide for all. In *Zombie Economics: How Dead Ideas Still Walk Among Us* economist John Quiggan exposes economic myths that keep getting buried and then raised again by blinkered, doctrine-driven economists.[28] Trickle-down economics is one of these zombie theories that needs to stay buried. What in fact happened after 1989 was that the rich got hugely richer and the gap between them and the poor (inequality) got bigger and bigger.

When the 2008 global financial crisis unfolded, a big collapse loomed. Massive changes in fortunes and welfare occurred as self-correcting feedback effects came into play – the classic kinds of re-organisation involved in maintaining the resilience of a complex system. The process inevitably involves reductions in some parts of the system, and in this case the short-term losses were deemed to be too great for them to play out, so governments and reserve banks propped up the existing, non-resilient structure of the system and bailed out those who had brought it to the brink. The growth paradigm survived. The pursuit of economic growth as the unquestioned goal of governments (on the assumption that it is essential for jobs), the failure of markets to include all that is important for us, and the rising levels of corruption all contribute to further declines in resilience and do not bode well for future generations. Bill Clinton's famous dictum 'It's the economy, stupid' needs rephrasing so people understand that, actually, 'It's the stupid economy' that needs fixing.

Real wealth

The heart of the growth problem is the focus on economic flows – GDP, how much money is flowing through the economy – rather than the stocks of our assets that produce the flows; and we've known for a long time that this is a problem. Robert Kennedy once said of gross national product (basically GDP plus income from abroad):

> Yet the gross national product does not allow for the health of our children, the quality of their education or the joy of their play. It does not include the beauty of our poetry or the strength of our marriages, the intelligence of our public debate or the integrity of our public officials. It measures neither our wit nor our courage, neither our wisdom nor our learning, neither our compassion nor our devotion to our country, it measures everything in short, except that which makes life worthwhile. And it can tell us everything about America except why we are proud that we are Americans.

The wealth of a country is its assets, its capital stocks, including natural capital (soil, water, forests), human capital (healthy, well-educated, skilled people) and built capital (roads, factories, dams, etc.). They are the asset base that delivers wellbeing, including such things as being able to breathe clean air, produce food, having a job, and the happiness that comes from experiencing a coral reef.

If the capital stocks decrease, then the flows from them diminish and so does wellbeing. If fish stocks are declining because of too much fishing, and harvests therefore also start declining, it makes sense to stop fishing for a while so the stocks can build up again. In real economics this is an investment, but using GDP as the measure of progress – how much money is flowing through the economy – it would be recorded as a loss.

The focus of governments should therefore be on building, or at least maintaining, the stocks of its wealth, but in virtually all of them it is on how to maximise GDP. There is little attention paid to the secondary consequences of doing this, yet in seeking to find the attributes that make systems resilient, studies of ecosystems and social-ecological systems have identified high levels of reserves, of all capital stocks, as being major contributors. Leading economists have developed an index called inclusive wealth,[29] which provides a framework for assessing real progress – but it's difficult to replace the well-established, simple but misleading GDP index. A useful step towards this has been the development of a GPI (genuine progress indicator), which takes into account all that GDP does but also includes the negative costs that growth is causing, such as pollution and increases in poverty rates. Interestingly, for many countries the changes in GDP and GPI were quite similar from the 1950s to the mid-1970s, but since then for most countries, especially the developed ones, GPI has levelled off or declined while GDP has continued to grow.

Homo sapiens (wise man) was a wise choice for the name of modern humans. It highlights the essential feature that distinguishes us from all other species: our ability to reason and reflect. But in the modern world the overwhelming dominance of neoclassical economics has led some, who actually do reason and reflect, to re-name us *Homo economicus*, meaning that everything people do is governed by economic rationalism to maximise our own economic wealth and to achieve the continued development of a bigger and more efficient economy. Real 'sapiens' economists demonstrate why this is wrong and why it is pushing us along a very undesirable and dangerous trajectory. As explained in an interview with one such economist from the Santa Fe Institute, Sam Bowles has shown experimentally that real people do not act as economic rationalists. They do not act only in terms of self-interest and personal benefits. For instance, they willingly forgo benefits due to them in order to punish cheaters. Their behaviour to achieve overall satisfaction is far more complex than economic self-interest, including many non-monetary factors. As Sam Bowles puts it, economists have shrunk real economics because the simplification allowed the 'mathematization' of economics, which doesn't explain how human beings actually behave. Adam Smith, the founder of modern economics, is usually only known for his laissez faire market ideas, the 'free hand of the market' as expounded in his famous 1776 opus *The Wealth of Nations*. But he was not that narrow and also believed that however selfish man might seem 'there are evidently some principles in his nature, which interest him in the fortune of others, and render their happiness necessary to him, though he derives nothing from it except the pleasure of seeing it'.

The job of the economy is to increase human wellbeing across society. If in some places and at some times it needs to grow for that to happen then it should grow. But if at other times under different circumstances the secondary consequences of growth lead to an overall decline in human wellbeing, or a great risk thereof, then it should not grow. What is wrong is the ideologically fixed focus on economic growth as the overriding requirement for the economy. In 1848 John Stuart Mill, one of the fathers of economics, stated: 'It is scarcely necessary to remark that a stationary condition of capital and population implies no stationary state of human improvement.'[30]

The mantra 'jobs and growth!' is driven by the belief that growth is necessary for jobs. But what kinds of jobs? In some countries there is a trend to reduce working hours and share more jobs: productivity per person is shown to increase. And the word 'jobs' is itself an obstacle. The rapid rise in technology is rendering many traditional industry jobs obsolete, and a fulfilling occupation is a better term. Self-employment is rising in most countries and for some it offers greater life satisfaction than a factory worker or office position. Should governments promote creativity in this area and encourage more of it? How can a government help to create quality of life independently of 'jobs'? Reducing inequality is a critical part of the solution. With wealth shared more equally, social capital rises and the demand for services, including new kinds of community help and involvement 'jobs', increases.

Inequality, power and social capital

In their book *The Spirit Level*,[31] Richard Wilkinson and Kate Pickett have developed an index of what they call 'social ills': the sum of things such as crime rates, numbers of people in jail, teenage pregnancies, obesity, and so forth. All these social ills cost society a great deal. To find out what it is about a society that determines the level of social ills, they first plotted the index against GDP per capita for all countries they could get data on. There was no relationship: it was a scattergram. But when they plotted the index against a measure of inequality (basically the gap between rich and poor) it was a straight line. The more unequal a society the higher its index of social ills. And here's what's really interesting: the cost *of* the social ills (inequality) is much higher than would be the cost of reducing inequality. The New Economics Foundation calculated that the costs of the UK's social ills from 2010 out to 2050 exceeded £45 trillion. So why doesn't reduction in inequality happen? Because power in a world driven by short-term profits is in the hands of those with allegiance to only their corporations and themselves: they have huge political influence and the inequality suits them.

In a complementary way to Wilson and Pickett's findings on social ills, Robert Putnam has explored the effects of social capital on a range of social 'goods'.[32] He obtained data for the USA from surveys such as 'do you trust other people?' and studies on such social attributes as the proportion of parents belonging to parent teacher associations, whether people attend local meetings, serve on committees of local organisations, give blood, donate to charity, and so on. And, importantly, in relation to the observation that a handshake is faster and cheaper than having to use a lawyer, he showed that over the first 70 years of the last century there were ~40 lawyers for every 10 000 employees in America but this number has started to increase as trust and social capital started to decline, so that by now the proportion of lawyers has more than doubled. He then combined all these attributes into a single measure of social capital for all the states and plotted them against measures of the sorts of things regarded as good in society. It showed a strong positive relationship with educational performance in schools, and in child welfare. Crime has a strong negative relationship: the higher the social capital the lower the crime rates, especially violent crimes. Good health is strongly positive. So is tolerance for gender, race and civil liberties. And in agreement with Wilson and Pickett's findings, there is a strong positive relationship between social capital and economic equality.

Combining Putnam's work and other studies, the attributes that stand out in determining social capital are high levels of trust, strong social networks and good leadership (see Chapter 7 for how these played out in Bikita's Chieftain areas). Further work is needed because more intangible attributes also appear to be important, such as empathy: having more people with a willingness and tendency to understand the views and needs of others, promoting cooperative and supportive behaviour.

'Inequality' and 'inequity' are both used to describe the difference between rich and poor, but social scientists explain that whereas inequality is just an objective measure of the difference, inequity has a connotation of unfairness. The poor end of society generally accept inequality when the unequal distribution does not seem distorted and unfair, but inequity is what people really resent and that gives rise to revolutions, as history has shown (let them eat cake).

As inequality rises, the disparity between the haves and have-nots gives rise to increasingly narcissistic behaviour on the part of the haves: their use of 'othering' intensifies. Drivers of more expensive cars are less likely to give way to pedestrians or other cars. The very rich have a greater sense of entitlement and are less generous, to the point where they are far more likely to take sweets from a bowl marked 'for children' (demonstrated in an experiment). This kind of arrogant, self-serving behaviour is part of the complex issue of power and the role it plays in societal development. Wilkinson and Pickett's final comment sums it up nicely: 'It is hard to avoid the conclusion that we become less nice people in more unequal societies. But we are less nice and less happy: greater inequality redoubles status anxiety, damaging our mental health and distorting our personalities – wherever we are on the social spectrum.'

With rising growth and inequality, behaviour changes, including a rise in corruption and, through exerting the power it brings them, the perpetrators defer and avoid justice, furthering the inequality. The aggregate wealth of the world's richest 1% of people is the same as the aggregate wealth of the other 99%. In the

USA, most Americans are unaware that the top 1% own 40% of the total wealth in the country and the bottom 80% own just 7%. It seems neither fair nor sustainable that the average CEO in America earns around 380 times what the average worker in the company earns. That smacks of inequity. The threshold between inequality and inequity has been crossed in several countries, and their susceptibility to shocks and disturbances, of all kinds, is increasing.

There is much that the understanding of long-evolved resilience in natural systems can contribute to helping understand and guide development in social systems. But sometimes a direct comparison is inappropriate and a comment is needed here on one such comparison.

The top-down effects in an economy are sometimes equated to the positive, diversifying effects of trophic cascades in nature. But this is inappropriate and misleading. To begin with, ecosystems cannot and do not need to keep growing. Once established, they are driven from their base of nutrient wealth and water, up through a fluctuating but essentially stable amount of vegetation and the trophic layers above to the top predators, whose activities diversify the layers below. Economic systems are designed to keep growing as though there are no limiting assets to constrain growth. The top 'predators' in economic systems do all they can to increase conversion up the food chain to them, reducing diversity below and therefore resilience. Only by continually growing can their demands be satisfied. Lions are not greedy. They kill and eat when hungry, and laze around in between. They don't keep accumulating kills as a sign of their success, to be admired by other lions – 'So how many wildebeests did you guys get last night? What, only five? We got eight.' They could no more use the extra seven than the super-rich can use and benefit from more than a fraction of what they have harvested.

Shifting to an economy for sustaining human wellbeing

The discussion so far has described the very real shortcomings of our present economic system, in terms of both its inherent unsustainability and unfairness. As we've seen, indicators of real progress, as opposed to the narrow and misleading GDP, have been developed but they have not yet been incorporated into the economic system. This brings us back to the third reason for not changing our behaviour, described at the beginning of this chapter: being trapped in a way of living from which it is very hard to escape. Change won't be easy but as more and more people become aware of and concerned about the need for it, hopeful developments are underway. I'll briefly describe two of them: one to do with the culture of corporations in the capitalist system, the other focused on changing the economic system itself.

The culture of corporations in the West is defined by four characteristics: pursuit of size and scale for market dominance; lobbying for regulatory and competitive advantage; advertising, with minimal or no ethical constraints; and lastly, use of borrowed funds to leverage investment. This began around 1920 and, although it powers on today, its world does not. Just for starters, around US$1 trillion per year is spent worldwide in subsidies to corporations favouring business as usual over more sustainable alternatives in industry. Fossil fuel subsidies alone amount to $650 billion a year. Most corporations aggressively oppose legislation against 'the free market', which is in fact anything but free: it is highly distorted in favour of corporate short-term profits.

All this is spelled out in the book *Corporation 2020* by Pavan Sukhdev.[33] He is a past head of Deutsche Bank's Global Emerging Markets Division and he makes it very clear that the world today demands a change to Corporation 2020, defined by four planks:

1. The disclosure of externalities (resource use, pollution and social impacts).
2. Limiting the amount of leverage: no more 'too big to fail' (as John Maynard Keynes put it: 'If I owe you a pound I have a problem, but if I owe you a million, the problem is yours'). Misuse of leverage was at the heart of at least the last four major financial crises.
3. Shifting taxation from taxing profits to taxing resource extraction (taxing 'bads', not 'goods').
4. Accountable advertising, including measures to avoid 'green-washing'.

Pavan is initiating new companies that conform to several principles, such as a cap on the ratio of the salary of the Chief Executive to the average worker, and making all employees shareholders in the company. The performance of these companies is promising.

The change from Corporation 1920 to 2020 is not a minor adaptive adjustment: it amounts to a transformational change and it will need both bottom-up and top-down support, from activist citizens, NGOs and visionary political leaders.

Changing the economic system to one that uses sustainable human wellbeing, rather than GDP growth, as its yardstick is the stated objective of a developing partnership of economists, ecologists and even some governments. Called the Wellbeing Economy Alliance (WE-All),[34] it is designed to facilitate the transformation from the present economy to a new economy that incorporates the real indicators of sustainable progress in human wellbeing, recognising that the economy is embedded in society and nature. It has a wide and growing membership and is developing an operational framework in a bottom-up way using these indicators, based on what is happening in various countries.

PART V

A way forward

11

Changing cultures

How can we transition to a different way of thinking, a norm that matches the reality of the biophysical world, and the reality that individual wellbeing is largely determined by the wellbeing of society?

In his famous 'to be, or not to be' soliloquy, Hamlet explores the option of continuing with the agonies of life in the situation he finds himself, versus the uncertain alternative of a peaceful life in the hereafter. It's a difficult choice for him to contemplate. There is a similarity between Hamlet's dilemma and the choice that confronts many who benefit from, and in many cases depend on, the social-ecological systems they live in. Their dilemma is whether and when a change in how their system is used might be necessary.

At the risk of stretching the analogy too far, resilience is about exploring the option of 'to change, or not to change': to continue with the existing costs and benefits of the way the system is being used, versus the uncertain costs and benefits of making a change. As an example at the global scale, the account of climate change at the end of Chapter 10 highlighted that the longer term costs of not doing something to curtail climate change will be much greater than the short-term costs of doing it. The longer the decision is put off the greater will be the costs, and they may well be irreversible.

Exploring the options at the scale of a community, a forest or agricultural region is not as clear as in the case of global climate change. However, the numbers on the short-term costs of making the change(s) are once again fairly easy to estimate, but the longer term costs of not changing necessarily involve making some assumptions, and so the estimates are given as likely ranges of values. The unwanted outcome of this is putting off the decision until it's too late, until the system has passed a threshold and can't get back to how it was. Without making the necessary change in time, the system can no longer stay the same kind of system.

This dilemma in deciding on where, when and how to change is a big question, probably the biggest question, facing the world as the global food–energy–water nexus looms: whether and how the social values and norms needed to make necessary changes can develop, in time, at national and international scales. There are signs of change in many countries and the hope is that the bottom-up influence of increasing social pressure on the short-term power/profit decisions of politicians and corporate leaders will speed up the process. Changes are happening through this bottom-up pressure, increasingly powered by the rapid spread of social media. There is also a dawning recognition at higher levels of the real economic value embodied in all the combined attributes of natural and human-used ecosystems, as opposed to just an immediate profit-making product.

In 2012 my daughter Kate spent a week in an ecotourism lodge in the Peruvian part of the Amazon, and as part of her eco-experience she engaged with a project on macaws and jaguars. Based on its magnificent biodiversity, ecotourism is now the fastest growing industry in Peru. Kate noted that the big problem in the Madre de Dios National Park when she was there was illegal gold mining, interfering with conservation programs. It was just 20 years earlier when the mayor of Yurimaguas, not that far from where Kate was, had discussed the future of the region, and the potential of the ecotourism industry was not even considered. I don't know if illegal gold mining occurred there at that time, but if it did it would not have been considered a conservation problem. Now, however, Yurimaguas does have a developing tourism industry and this was possible because the forests in the region still had the potential for it. The change from just agro-forestry to an economy with a significant, rapidly growing ecotourism component was made in time, before the option was lost. And though it is tourism that secured the forests, their benefits to the people in the region extend well beyond that.

The value of the tourism industry is just one of the services rainforests provide. They also provide pollinators, pest control, carbon sequestration, regulation of water supplies and others. Getting accountability for the value of all these services into mainstream economics and on to government agendas would make the world a lot more sustainable and a lot more equitable. The difficulty lies in getting agreement across levels in society on what is valued and what the forests can provide. This was well illustrated for me on a Saturday morning I spent in the company of some 30 young Malaysian schoolchildren on the nature trail at the Forest Research Institute of Malaysia. I was visiting the Institute and took the opportunity to join in with the school tour.

Tropical rainforests are hot sticky places and at our second stop I was gazing up into the canopy when I felt a welcome breath of air fanning the back of my neck.

Two little girls linked arm-in-arm in best-friend fashion, each with a fan in hand, had decided to favour me. I'd heard them whispering and giggling behind me and then one of them had begun to fan me while the other fanned the two of them. Out of such stuff are friendships born. The three of us formed a team to search for particular fallen leaves, as instructed by our leader, and dutifully smelled the resin oozing from a camphor tree. We examined ants, tested the ability of leeches to locate our warm skins, gazed in awe up into the canopies of huge trees and were told about which of them produced fruit of different kinds at different times (I was getting a piecemeal English translation from the leader).

If it were up to those children and their teachers and that forestry officer, the future of Malaysia's forests would be secure. Their use would be tempered by knowledge of what's needed to ensure forest regeneration and the continued supply of all its goods and services, including timber. I left them after sharing a picnic lunch with my two partners while they pondered the question of why some of the trees had buttresses and others did not.

But the decisions about Malaysia's forests are not made at that local scale. How the forest works, and the range of forest values and services being inculcated into the children, so far enters little into the equation that determines government logging policies. At that scale the equation is dominated by short-term economic growth and industry demands for access to the timber. Only if the decision makers take a long-term view that incorporates the real value of all the goods and services the forests provide, and will continue to provide into the future, will Malaysia's forest assets be secure. And should Malaysia choose that option, it will be more resilient to whatever it has to face in the future.

Not long after my excursion with the school children, I visited a Dayak village in the rainforest of East Kalimantan on the island of Borneo. The Indonesian Government had recently devolved ownership of the forest to the communities living in them, and the responses of different communities to their newfound powers varied. Many sold their forests to multinational logging companies, some simply for short-term gain but for others it seemed to be a distrust that they would continue to have ownership in the future: better to grab what they could now before a new government takes it back.

The village of Setulang, however, had refused logging company offers and had opted to keep its forests for its own use. On the flatter areas, which had been cleared, the villagers practised rice rotation. They had worked out a system of selective tree harvesting in the hilly parts and had traditional restricted forest areas where they took us to show the many products and services it provided them. One of them, which illustrates the range of products they valued, still sticks in my mind. We were billeted, in pairs, with various families and I was fortunate in being with a Dutch colleague who spoke Indonesian, and our host could speak a little of it. His wife spoke only Dayak. On the evening we arrived he insisted we join him in having a drink. He produced a large jar in which there was a whole deer fetus in some kind of rice liquor. It had a weird astringent taste, but he assured us that drinking it would render us incredibly fertile.

The village was under pressure, with logging and corrupt deals always on offer, but up to that point the villagers had evinced a high level of social capital. Another example of a long-enduring traditional system of using natural resources. Times change. They now have a growing experiential tourism operation, and with their cooperative behaviour it will hopefully persist.

It takes a long time for such systems to evolve at local scales. It's much harder to achieve at national scales, but with the knowledge of the kinds of 'rules' that are needed the process can be aided by national governments and sub-national governmental bodies. It will be much more difficult at the global scale.

Norms and ethics

The choice about whether or not to smoke began as an individual choice, and when the numbers of non-smokers and those objecting to smoking reached a tipping point it became a social no-no and the new societal norm then had a strong feedback effect on smokers, leading to new rules and regulations. Non-smoking became a new social norm. When it comes to difficult decisions about competing values, social norms provide a reference touchstone that can lead to a change in what is considered acceptable behaviour: what's right and what's wrong. As they strengthen, they become ethics: the rules for behaviour.

In our world today, a critical societal ethic that needs to advance very quickly revolves around how we use natural resources. Back in 1949, wildlife ecologist Aldo Leopold expressed this eloquently in his classic book *A Sand County Almanac*:

> *When god-like Odysseus returned from the wars in Troy, he hanged all on one rope a dozen slave-girls of his house-hold whom he suspected of misbehaviour during his absence. This hanging involved no question of propriety. The girls were property. The disposal of property was then, as now, a matter of expediency, not of right and wrong. Concepts of right and wrong were not lacking from Odysseus' Greece: witness the fidelity of his wife through the long years before at last his black-prowed galleys clove the wine-dark seas for home. The ethical structure of that day covered wives, but had not yet been extended to human chattels. During the three thousand years which have since elapsed, ethical criteria have been extended to many fields of conduct, with corresponding shrinkages in those judged by expediency only ... There is as yet no ethic dealing with man's relation to land*

and to the animals and plants which grow upon it. Land, like Odysseus'
slave-girls, is still property. The land-relation is still strictly economic,
entailing privileges but not obligations.[35]

The land ethic has come quite a long way since Leopold's day, including water, plants and animals, and to some extent even soils, but it needs to evolve a lot more rapidly. What is appealing in Leopold's analysis is his clear exposition of how ethics evolve. He presented the land ethic as a product of social evolution because, in his words:

Nothing so important as an ethic is ever written. Only the most superficial
student of history supposes that Moses 'wrote' the Decalogue; it evolved in the
minds of a thinking community, and Moses wrote a tentative summary of it
for a 'seminar'. I say tentative because evolution never stops.

The resource/environment ethic has a way to go before it is a sufficient counter force to the argument that economic growth is essential for quality of life, and that increased resource use is essential for economic growth (GDP). But one doesn't have to go far back to refute the argument that unacceptable behaviour is necessary for the good of the economy. Some 19th century economists used it in favour of retaining the slave trade, claiming it was an essential element of a modern economic system. The fact that it wasn't became apparent soon after it was abolished. The transportation of 11 million slaves over 300 years at an incalculable cost in human misery may have given a boost to the early development of American agriculture and made some people wealthy, but it was not essential for the functioning of the economic system, nor for the improvement of quality of life, nor for the progress of humanity. Likewise, destroying the natural resources of the planet is not essential for real economic development. Neither is the continued use of carbon fuels for energy production and, in fact, although once again a few people are making big short-term profits out of it, overall it is bad for business. Rising levels of obesity have triggered a hot debate in Australia about introducing a tax on sugar. The bulk of society and the medical profession are for the tax; the cane farmers, sugar industry and their representatives in government are strongly opposed.

The smoking and slavery histories, and the current machinations of the fossil fuel and (at least in Australia) sugar industries, illustrate that ethical influences operate in two connected ways. First, in a top-down way through laws that control the actions of individuals and companies (such as pollution and tax laws) and, second, through the bottom-up influences of individual behaviour and choices. Responding to demands from consumers, the rise of ethical investment companies gives consumers a choice to invest their money in renewable energy companies, for example, and as that happens it feeds back both to established energy companies trying to resist change and to new ones looking for opportunities. Sometimes governments take note of this through assistance packages for new, developing industries, but all too often they cave in to the demands of and political party contributions from big established business trying to resist change. However, encouraging signs are underway. The bankruptcy of several of America's largest coal companies, including the largest (Peabody Energy Corporation) shows that more and more people do not want to invest in coal and choose instead investment in renewable energy. Peabody's share price dropped from over $1000 in 2011 to just over $2 in early 2016.

It is in this bottom-up process of social, or cultural, evolution where hope for the future lies. It takes many generations for biological evolution to produce a change in a population, but social/cultural evolution is different and it can be very fast. It involves what Richard Dawkins calls 'memes': ideas that are passed from one individual to others, not just *an* other, and not just vertically through reproduction, but sideways in all sorts of ways. With the rapid development of social media, this is becoming increasingly fast and increasingly widespread. The development of memes is about the evolution of norms: the idea catches on and conformity with it grows. And, as the bottom-up influence strengthens, it leads to a developing understanding of why it's considered important, a developing ethic, and an emerging narrative that becomes a part of social discourse that influences others.

Our choices influence other people's behaviour and add up to the development of cultural norms, and normative behaviour is the expression of a society's culture, eloquently described in Cristina Bicchieri's *The Grammar of Society*:

> Norms are the language a society speaks, the embodiment of its values and collective desires, the secure guide in the uncertain lands we all traverse, the common practices that hold human groups together. The norms I am talking about are not written and codified; you cannot find them in books or be explicitly told about them at the outset of your immersion in a foreign culture. We learn such rules and practices by observing others and solidify our grasp through a long process of trial and error. I call social norms the grammar of society because, like a collection of linguistic rules that are implicit in a language and define it, social norms are implicit in the operations of a society and make it what it is. Like a grammar, a system of norms specifies what is acceptable and what is not in a social group. And analogously to a grammar, a system of norms is not the product of human design and planning.[36]

The persistence of the change from smoking to non-smoking is what makes it a change in norms. Some societal changes are trivial and very fast, such as fads and fashions, and do not reflect a change in norms. Norms have significant effects and it may take a long time for them to change, such as slavery and foot-binding in China. Though it was clearly negative in its effects, foot-binding persisted for a thousand years despite being banned a couple of times by emperors. No matter what parents thought of it, if they wanted their daughter to marry into their level of society the norm was that she had to have her feet bound. When it ended, however, it did so within one generation, initiated by antifoot-binding societies whose members pledged their daughters' feet would not be bound and that their sons would only marry women without bound feet. Because there was already widespread sympathy with the idea, the tipping point was at a very small proportion of the population.

A problematic norm underlying the increasing size of the world's human population and its impact on the environment is the desired/expected number of children in a family. If most of the families in your society have six or eight children then that's the norm you will aspire to. As the number declines it reaches a tipping point where the new norm is likely to be several less, and most will fit in with the new norm – as has happened in several developing world countries. But there are many factors that play a role in whether and where such a tipping point might happen, economic uncertainty and religion being two prominent ones. Recent surveys of women around the world have shown that if all unwanted babies could be prevented there would be global zero population growth. However, it's hard to get into effect. The most effective contraceptive device for women in rural areas is an inter-uterine device, but some religions forbid it: a very hard norm to change.

Apart from environmental consequences, as people become increasingly concerned about the homogenising effects of globalisation there is an increasing desire to retain and express their own identities, seen in expressions of local and national values. As the size and development of the 'community' to which we are deemed to belong increases, so does the need to somehow affirm our own identity. A newspaper headline on the day of the 2014 Scottish referendum on independence stated 'It comes down to livelihood *v.* identity'. The Brexit vote that took Britain out of the European Union is part of it and, in discussing the 'illegal' Catalan independence vote in October of 2017, a Spanish economics professor in the University of Edinburgh wrote that: 'Economics is a sideshow – everything is about identity politics. It's a definition of "us".' Of course, the identity of an individual, a society, a nation, is never constant. It evolves, adapting to a changing environment: just think about the identities of various medieval societies and those same societies today. But if the changes in a person's or a society's environment are too big and/or too fast the person or the society is threatened and reacts negatively. As individuals, our identities are expressed through interacting with and feeling connected to a limited number of other individuals. We don't want to be just a part of some amorphous mass, and there is an interesting biological basis for this.

British anthropologist Robin Dunbar plotted the average group size of the 39 different kinds of primates against their brain size. It turned out to be pretty much a straight line: the bigger the brain, the bigger the group size. And when he extended the line to the brain size of humans the predicted average group size is 148 – let's say 150 (now known as 'Dunbar's number'). Some supporting evidence for it comes from Neolithic farming villages, which numbered around 150.

What keeps primates together is grooming: interacting and developing trust, and the bigger the group size the more time that takes. It's not too bad if there are only a score or so individuals but, based on the proportion of time primates spend on it, it works out that for 150 of them 42% of their time would have to be devoted to social grooming. One suggestion that came from this is that language arose as a 'cheap' means of grooming, cutting down the time it takes and thus allowing humans to get on and do other things together, and so allowing the bigger group size.

How does this relate to modern group sizes? A recent study indicates that 150 is about the maximum size for Twitter conversations. How many people do you maintain contact with? Most people initially think 150 is on the high side, but if you take into account the 'network of networks' structure in society it extends the number and so networks take on special significance, as do shadow networks; no 'grooming' needed but they can be reinstated quickly when required.

Think about your own networks: work, family, play. Each has levels (in network theory the strength of the connections) and the total easily adds up to quite a large number. It's all those connections that give your life meaning and purpose, that create your identity – and it has little to do with economic rationality.

How will the growing tension between global interconnectedness and the need for our own identity play out? Will the global corporations prove too powerful, or will the rising individual unease coupled with the power of social media and ethical investment choices provide a sufficiently strong feedback to the global corporations? And what roles do national governments play in this? At present, most of the democratic ones appear to be following the wishes of the corporate world.

Our future wellbeing depends on whether or not there is a change in culture around the world to a norm that matches the reality of the biophysical world, and the reality that individual wellbeing is largely determined by the wellbeing of society. A change in focus from 'me' to 'we'. Can a bottom-up cultural shift induce, in time, a change from Corporation 1920 to Corporation 2020?

12

A resilience pathway

Given a goal of long-term human wellbeing, under its mantle of coping with uncertainty, resilience gives priority to avoiding unacceptable pathways, and we learn to ride nature piggyback rather than trying to dominate her.

Through chance with wisdom

Three young guys were walking along a road when a man approached and asked if they had seen his camel. They hadn't, but because they were clever and observant they were able to tell him that indeed a camel had been there, was blind in one eye, lame in one foot, had been carrying honey on one side, butter on the other and also a pregnant woman. The camel driver then had them arrested on the grounds that, because all these things were true, they must have stolen his camel. That's the fable about the three princes of Serendip, which goes on to show how these princes were able to deduce the things they did by careful observations, by their sagacity. Based on the fable, Thomas Walpole coined the word serendipity in 1754, to mean 'making discoveries by accidents and sagacity' and it's the combination of sagacity (wisdom/discernment) and accidents (chance) that makes it fit well with resilience.

Chance plays a significant role in how ecosystems and social systems change, in unpredictable and unexpected ways. And this is not just characteristic of them, it is essential for their continued wellbeing. It is what distinguishes them from physical systems in which random changes are regarded by physicists as background noise, a nuisance. In the sense of the many things we care about in them, with all their complexity, social-ecological systems just don't behave like simpler physical systems. It is the chance variations that maintain their diversity, keep their options open and allow new ways of doing things to keep them functioning in the same kind of way. Remember, what's noise to the physicist is music to the ecologist, such as the chance blunderings of hippos that keep the Okavango swamps in an ever-changing, healthy state (Chapter 3). Embracing chance events and learning to use them in wise ways is a key part of practising resilience.

Years ago, Mark Westoby, Imanuel Noy-Meir (two good friends) and I developed an approach for how to do this: how to apply resilience in practice, by thinking about rangelands (in our example) as being able to exist in several different states with conditions and chance events that brought about transitions between them. In managing such a system, being able to absorb shocks and to change when needed, the managers would see themselves as facing an oncoming stream of events, a mixture of opportunities and hazards, and their objective would be to seize the opportunities and evade the hazards, as far as possible.

Just how you do this depends on where you come from: the worlds of psychiatry, engineering, economics, ecology and others, even architecture. There are more than 70 definitions of resilience and though the details differ, in most of them the basic precepts are much the same.

Homage to St Publius

One of the more famous papers in ecology has the intriguing title 'Homage to Santa Rosalia' with the subtitle 'Or why are there so many kinds of species?' It was written in 1959 by George Evelyn Hutchinson, Professor of Ecology at Yale University, who worked on competition between closely related species. He had been puzzling over the question of how such species managed to co-exist. If they were so similar, why didn't one of them out-compete the other and drive it to extinction? While looking at a small pond in which there were two species of water beetles, he happened upon an idea that later became known as 'Hutchinson's ratio'.

One of the species was larger than the other and had just finished breeding and the smaller one was about to start. Watching them swimming around together set him wondering about why there were two species in the pond, and why only two and not 20 or 200. And this led him to wonder about how much difference in size and behaviour was necessary to avoid them being too similar to co-exist. So he got hold of lots of data for other co-existing, closely related species (such as Darwin's finches on the Galapagos) and came up with the ratio of 1.3. That is, one of the species had to be at least 1.3 times bigger than the other for them to co-exist.

The small pond Hutchinson was looking into when he had this inspiration was just below a cave near the summit of Mount Pellegrino, in Sicily. In front of the cave is an old church dedicated to Santa Rosalia, who died in the 12th century. Her stalactite-encrusted skeleton was discovered in the 16th century and she became the patron saint of Palermo. Hutchinson's homage to her was for his idea about how different two species had to be for them to continue living together.

Forty years later, not far from Santa Rosalia's church, on the Maltese island of Gozo, I was with a group of colleagues pondering the question about what different kinds of species *do* when living together. Are they all 'needed'? What functions do they perform? What would happen to the ecosystem if some of them were removed?

It turns out species have two kinds of roles. One is they do different things: they belong to different functional groups, such as predators, grazers, seed dispersers, pollinators, nitrogen fixers and litter decomposers. They are all needed for the ecosystem to keep functioning. Within each functional group there are generally several different species all performing the same function but in different ways and with different responses to the environment. And this is the second role that differences between species play – provide a diversity of responses to the different kinds of disturbances an ecosystem has to contend with. It is this response diversity that makes ecosystems resilient to disturbances. An ecosystem with just one legume species may continue to fix the nitrogen it needs as long as nothing happens to it. But if a drought or fire or a disease eliminates that legume then the function of nitrogen fixation is lost and the ecosystem will change radically. However, if the ecosystem had several different legume species, with different responses to environmental shocks – some better at surviving drought, others able to survive frost – it would be able to continue fixing nitrogen whatever environmental shocks it receives.

The disaster in Shepparton's milk industry (Chapter 8), when millions of litres of milk had to be poured onto fields, was due to a lack of response diversity.

All too often the economic growth norm regards response diversity as costly, inefficient redundancy and gets rid of it: all you need is the one, best performing way of doing it. In some cases, however, where the consequences of failure are really high, the risk is recognised and redundancy is built in. Most passenger aircraft, for example, have more than one operating system installed. Some are 'quadruplexed': they have four independent systems in case one or more of them fails.

At the end of the meeting on Gozo, we took the ferry back to the main island of Malta and got caught up in a procession through the streets of the town of Floriana, celebrating the feast of Saint Publius, the first Bishop of Malta, who sheltered St Paul when he was shipwrecked there. He was martyred by Emperor Trajan in AD 112. As we walked along behind the float sporting the effigy of St Publius, I was thinking about the difference between *how* species manage to live together (Hutchinson's puzzle), and *why* they do (our puzzle). It struck me that it wasn't far from here that Hutchinson had come up with his idea of the how. So I thought it appropriate to pay homage to St Publius for the idea about different kinds of diversity, and why it is so important to have different kinds of species to perform the same function: response diversity.

This functional role of diversity was the focus of my own research in Australia, using rangelands as the 'system'. I owe much for the ideas that came out of this work to the great group of scientists with whom I interacted at meetings like the one on Gozo: meetings of the Resilience Alliance. Founded by Buzz Holling, it arose out of a group of like-minded natural and social scientists from a mix of backgrounds grappling with similar problems. It operates on 'island rules', which in short means meeting on small islands, taking long lunch breaks for small group discussions, getting together over drinks at sundown, and no after-dinner lectures. The rules extend to who gets invited to a meeting. In suggesting someone as a potential participant, you needed to ask the question 'Is s/he good on islands? Would you like to spend a week on a small island with this person?' People from quite different backgrounds willing to discuss how different ideas could come together in a bigger picture were welcome. Those who wanted only to tell you about how they thought the world worked were not.

Sometimes we redefined what an island was. One of our early meetings took place in Malilangwe, a wildlife reserve in south-east Zimbabwe – an island of nature in a sea of agriculture. On our first evening in camp we sat around an open fire and were served peanuts and biltong: the South African equivalent of jerky, but spicier. Everyone knew of it and most liked it. But there was also another dish of fairly crisp and crunchy delicious things about the size of a large worm, which is what they were: dried mopane worms. It's a local delicacy in southern Africa wherever there are mopane leaves, and the worms are very nutritious. In some years they are hard to find and in others they are super-abundant. All in the group were enjoying them until they found out what they were. Some took a close look with a torch, saw the eyes and dried head (the crunchy part) and nearly retched. How variable, intermittent supplies of ecosystem services, such as this mopane one, change under changing land use was part of what we were exploring.

In that meeting we spent a morning with a group of local game ranchers who told us about the region and what was happening, and how they had given up cattle ranching and amalgamated their properties to run as larger wildlife game viewing and hunting areas. They did this after a bad drought killed ~90% of cattle in the area but much of the wildlife had survived. Asked why they would take such a chance not knowing if it would work, their reply was they had no insurance in Zimbabwe, there was no chance of any help from government or any other possible back up and so they had to take transformative action. Had there been the kind of drought relief that is customary in the USA and Australia they would no doubt have continued cattle ranching. They would have been given help not to change, but would never have received help *to* change. For these ranchers, their wildlife safari business turned out to be far superior to cattle ranching in terms of both the state of the land and their profits.

The island meetings compared how different ecological and socio-economic systems are connected and how they change, often in unexpected ways. Out of them the ideas of social-ecological resilience were formulated. Describing all these ideas and the examples behind them needs a textbook (in fact several have been written) and that kind of detail doesn't belong here. A short synopsis, however, helps understand how we might find a way through the impasse the world is in.

The elements of resilience

The characteristic feature of resilience in any system is that there are limits to how much it can change and yet still function in the same way: still be the same system. A few critical variables determine the dynamics of all the others and there are limits to how much these variables can change before the whole system starts behaving differently. Once the threshold amount (tipping point) is passed, instead of tending back towards how it was, the system starts changing in a different direction into some other 'state' of the system.

Often this other state is considered 'bad' and the objective is not to cross into it. But sometimes, as with the *Salvinia* weed, it is desirable to cross it in order to get back to the kind of system it was, and the objective is to work out how. Therefore, though the alternate states of a system may be considered good or bad, resilience itself is neither. It is a property of a system and both the good and bad states may have high or low resilience.

Resilience is not the ability to 'bounce back'. It is the ability to absorb a disturbance and re-organise so as to keep functioning in the same kind of way. To emphasise this, resilience is not about not changing. Changing and re-organising in response to disturbances is necessary for maintaining and building resilience.

In order to understand resilience you need to think about whatever it is you're interested in as a system, which means working out how all the important bits are connected, how a change in one might lead to changes in others, and how these changes might lead to further changes. It is these unexpected secondary effects that mostly cause systems to cross thresholds and start behaving differently.

The single biggest resilience mistake people make in managing any system – a person, an agro-ecosystem, a city – is to focus on only one scale, the scale of immediate concern, the focal scale. All systems function at multiple scales and the cross-scale effects are at the core of understanding the resilience of the whole system. Failure to understand and deal with this is a prime reason behind the failure of selective growth policies.

Focusing on the resilience of a particular part of a system – the resilience 'of' something 'to' something – is about learning how to avoid crossing some particular threshold as the system changes in response to a particular kind of disturbance, and though this might make the system resilient in that way, it can reduce its resilience in other ways. Furthermore, knowledge of these other ways, of other thresholds, often doesn't exist, but, knowing that other thresholds are very likely to exist, how can you make the system generally resilient in all ways to all kinds of disturbances? What are the features of a system that make them generally resilient?

Probably top of the list is diversity, and in particular response diversity, described above: don't put all your eggs in one basket. An agricultural region that produces and relies on one crop is in danger of losing all its food/exports if something, such as a disease, destroys that crop.

The problem is that very often the loss of response diversity isn't noticed until it is too late.

Perhaps second on the list is having reserves, such as seed banks and the underground reserves in Hwange's sandy ecosystems, and financial savings and memory in social systems. When the 2004 tsunami hit the coasts of Indonesia and Thailand, the sea was at first sucked out, away from the shore, by the development of the massive wave. Those who walked down following it, intrigued by what was happening, were swept away and drowned. But in some of the islands the memory remained that 'When the sea runs out, you run uphill'.

Another important attribute is being able to respond quickly to shocks and to changes in the system. This has evolved naturally over time in ecological systems, but not in social systems: having too many steps in a reporting and approval procedure significantly slows down response time. The widespread trend of more and more checks and balances to promote cautious, safe operating procedures – including safe in the sense of avoiding litigation procedures – reduces nimbleness and resilience.

A difficult attribute to get right is the amount of connectivity. An ecosystem, a city, or any complex system that is fully connected in all ways is very vulnerable if something bad gets in: a disease, a fire, a bad idea. This can spread rapidly through the whole system, giving little time to deal with it. On the other hand, having too few connections slows down responses too much and prevents the system from being able to self-organise and change when needed. Being relatively modular – having loosely connected sub-units (modules) that have tight internal connections – confers resilience. But how modular? What's the optimum amount of connectivity? The kind and degree of modularity needs to change as the system and the external environment changes. Cities, for example, develop in a modular way – different town councils, and so on – and they can become too rigid, and non-resilient, if prevented from adaptive change. It's another Goldilocks question and, rather than trying to determine the optimum kind and amount of modularity, perhaps the best that can be done is to determine, for the system at this time, where connectivity may be too high or too low, and where and how changes might be made.

A related connectivity problem emerges from concern about resilience of the whole system versus parts of it. The blackout across Italy (Chapter 8) was due to a doubly fully connected system of the power stations and the internet of computers controlling them. The mass blackout in New York after Hurricane Sandy showed that unwillingness to risk a shutdown in one part of the city, keeping the city fully connected, made the whole system riskier and less resilient than would have been the case if the individual parts had been riskier.

In contrast to the internal connectivity question is the issue of external connections, and the evidence says that an 'open' system is more resilient than a 'closed' one. In broad terms this means allowing emigration and immigration: bringing new genes into species populations, new species from surrounding ecosystems (not exotic species from other continents), new ideas and skills in social systems. In other words, bringing in new kinds of both functional and response diversity.

In social systems, a particularly important resilience attribute is having high levels of social capital, enabling them to be flexible and adaptable, able to absorb unexpected shocks and able to embrace and use failure as a way to learn, which is probably most often achieved by probing the boundaries of resilience. As discussed in Chapter 6, this is called 'stress inoculation' in psychology and requires strong connections in two ways: a support system of trusted, positive people and you helping and being responsible for others. The three attributes that stand out in determining social capital are high levels of trust, strong social networks and good leadership.

In an earlier book about resilience,[37] after listing the attributes we had distilled from the literature, my co-author and I asked readers to let us know if there were any we had missed. We received quite a lot of replies, and they fell into two areas: empathy/humility and fairness/equity. Since then, the more I have engaged in resilience assessment meetings the more I have come to realise how important, yet how unrecognised, these two attributes are.

In summing up his book *Fragile Dominion*,[38] Simon Levin lays out eight 'commandments of environmental management'. They largely overlap with the elements above but two points are particularly noteworthy. The first is in the complementarity of his first two commandments: 'reduce uncertainty' and 'expect surprise'. Do whatever you can to learn about your system, spread risks and diversify, but know that you are never going to get it all right so expect and prepare for surprise. The second point is in his final dictum 'Do unto others as you would have them do unto you', in which he makes the crucial point that societies can only survive when there is action for the collective good: 'we' versus 'me'.

These elements of resilience have mostly come from studies on social-ecological systems, but other disciplines have come up with their own. In the political science and sociology arena, Aaron Wildavski came up with six resilience principles – again overlapping with those above:

Homeostasis: Systems are maintained by feedbacks between component parts that signal changes and can enable learning. Resilience is enhanced when feedbacks are transmitted effectively.

Omnivory: External shocks are mitigated by diversifying resource requirements – failures to get or distribute a resource can then be compensated for by alternatives.

High flux: The faster the movement of resources through a system, the more resources will be available at any given time to help cope with perturbation.

Flatness: Overly hierarchical systems are less flexible and hence less able to cope with surprise. Top-heavy systems will be less resilient.

Buffering: A system that has a capacity in excess of its needs can draw on it in times of need, and so is more resilient.

Redundancy: A degree of overlapping function permits the system to change by allowing vital functions to continue while formerly redundant elements take on new functions.

This final element of Wildavski's set of principles emphasises a constant tension between trying to maintain resilience and trying to get it into some perceived optimal state. Resilience calls for some messiness in how the system works and what it looks like, as expressed by A.C. Grayling in his comments about democracy, and nicely captured in the last line of Pamela Brown's poem 'Missing Up': 'I like to keep some mistakes in; like drips in a monochrome painting.'[39]

Adopting a resilience approach for a particular state of a system might well meet current objectives and needs, but as the world changes so the context of your system changes, and sometimes what seemed to be the right kind of system becomes a problem. Mostly this is resolved through piecemeal adaptive changes that enable evolution of the system so that it continues to be what is wanted. But sometimes the external changes are of a kind, and have a speed and magnitude, that exceed the system's adaptive capacity. In this case, to avoid being stuck in what is now an undesirable system, there is a need for radical change: transformational change to a different kind of system. Being able to do this is also part of being resilient.

Transformational change

In a meeting run by the Global Environment Fund on how to increase resilience of agricultural systems in the developing world, a young man from sub-Saharan

Africa made an impassioned plea: 'We don't want to make them more resilient! They've failed, and can't give us what we need. Things are just getting worse and we have to change what we're doing. We have to find some different way of living. That's what we need help for.' His plea highlighted the problem of many aid programs: help not to change instead of help to change. He was calling for help to transform from the present failing system to something different that could deliver better human wellbeing.

Being able to transform – transformability – is still part of resilience because it depends on what you want to make resilient. If a system has crossed a threshold into something bad, or it's inevitable that in the way things are going it will do so, and it isn't possible to cross back again, then it is time to transform into some other kind of system that can deliver what is wanted. The Australian city of Wollongong originally developed as a mining port, but went into decline as market and economic changes reduced this activity. It is now transforming into a centre of higher education with emphasis on a top-class university, a centre of fine arts, tourism and eco-friendly electricity. The famous cod fishery of Newfoundland collapsed in the 1980s due to a combination of over-fishing and environmental change. People migrated out and it has now transformed into a fishery of snow crabs and shrimps, with different kinds of equipment and people: a different identity. Understanding and dealing with transformational change is now at the forefront of resilience science and practice: knowing when, where and how to become a different kind of system, or to avoid being transformed into one that isn't wanted.

Transformation and resilience can interact. Transforming in one place can increase resilience in another, and this can involve trade-offs across scales. There isn't enough water in the rivers of the Murray–Darling Basin, Australia's food bowl, to keep all the existing irrigation systems going. Irrigation licenses were issued in a very wet phase around 70 years ago and trying to keep them all going as well as keeping the rivers healthy is reducing the resilience of all of them. So some need to transform into some other kind of agricultural or even non-agricultural system in order to increase the resilience of the remaining irrigation systems, and thereby ensure a viable and sustainable irrigated agriculture industry across the whole Basin. But no bureaucrat or politician wants to go out on a limb and say 'Sorry folks, you are the ones who have to give up your water and do something else so everyone else can keep going.'

As the global environment changes, there are more and more instances where the option to transform is no longer possible. Sadly, if you haven't had the pleasure of snorkelling over a coral reef you should try to do it fairly soon. To begin with, corals like clean, clear water with not many nutrients. They and the tiny algae that live inside them thrive and grow, but when nutrients increase through runoff from agriculture on adjacent land the reefs favour large, leafy algae that grow on rocks or coral remnants. Whenever there is a bare area, before new little coral polyps can get established, algae grow vigorously. I described earlier how fish that graze on algae help control this up to a point, and how fishing pressure and the trophic cascade effect is giving the algae an advantage. The combination of fishing pressure and added nutrients has been increasing over many decades now and many reefs in the Caribbean and Pacific are now ex-coral reefs, converted into beds of algae.

This coral reef conversion has been a very undesirable transformation, but it's one that could be reversed by stopping runoff and fishing. However, a new major threat has emerged: coral bleaching. When the water around corals gets above a certain temperature the tiny algae inside them, supplying them with the energy they need, are expelled or abandon the corals, which then lose their pigments and die, leaving expanses of white, dead corals. Such bleaching events have occurred in the past, but there was a long time between them: enough for new corals to establish. Enter global warming, and as the oceans have heated up all this has changed. Since the first bleaching event was recorded in the 1980s the Great Barrier Reef in Australia has experienced four more: in 1998, 2002, 2016 and back-to-back with this, in 2017.

As warming increases, the return time of bleaching events will get shorter and shorter, always less than the time needed for corals to recover, so the reefs as we know them will not, cannot, survive. Stopping runoff and fishing will increase the resilience of the corals and delay the collapse for a while, but it cannot stop the process. Some deep corals will persist and recent studies show that some corals are more heat tolerant than others. Also, coral polyp migration may lead to some corals establishing on rocky reefs in cooler southern waters. There is already active research into this in Australia's Coral Reefs Centre of Excellence, aimed at developing a transformational pathway to some new kind of 'coral' reef system that is something the societies who live around and depend on coral reefs will value, and that will retain as much of the existing coral reef system as possible.

In the irrevocably transformed ecosystems of Hawai'i (Chapter 9), the objective is now to guide a transition out of the unwanted replacements to new kinds of ecosystems people will like, and that will retain as many of the remaining native species as possible.

On the other side of the world to the tropical coral reefs, a transformational change is happening in the Arctic Ocean in response to global warming. The decline in ice cover and warmer water are leading to changes in fish populations. Local people report an increase in salmon species moving up rivers to spawn. There are five species of salmon in the Arctic, plus many other kinds of fish, and there may be increases in salmon coming in from the Atlantic and Pacific oceans. It makes sense to protect existing and potential spawning grounds in the rivers in the face of increased Arctic region development, to try to guide a transformation to a new, different kind of fishing industry that indigenous people depend on.

The coral reef, Hawaiian and Arctic fishery examples are all cases of transformational changes that are underway and can't be stopped, so the need is to try to guide the process. But in much of the world today the problem is how to reduce the resilience of a system that is failing to meet people's needs in order to facilitate transformation to one that will enhance human wellbeing.

Intentional transformation is not easy. Nobody likes fundamental change and one of the hardest parts of transforming is getting past the resistance to it, past the state of denial that it's necessary. But it's a case of the first rule of holes – when you're in one, stop digging. Find or create something else to do. Sometimes it takes a crisis for this to happen.

Cycles of gridlock and renewal

A new forest starting out on a bare area grows rapidly with an abundance of space and available resources. As more species arrive, this growth phase becomes more complex, eventually reaching a mature stage where net growth is zero and more and more organic matter is bound up in wood or litter. Such 'old-growth' forests are much valued for their conservation value, with a host of plant, mammal, reptile, bird and insect species that only occur in them. Hollows that develop in old trees, for instance, provide nests for birds and arboreal mammals. Conservationists do their best to protect these old-growth forests, which are becoming increasingly rare.

A new business starts out as a growing, nimble organisation able to make quick decisions and changes but inevitably reaches a phase with more and more connections, rules and 'dead' matter. It becomes less and less able to respond to and successfully deal with external shocks.

In all kinds of systems, as the growth phase tails off they enter into what is called a conservation phase. They becomes more structured, conserved and interconnected, becoming more rigid and moribund the longer they stay in it. This can develop into a 'rigidity trap' that is very hard to escape. Lance Gunderson has been studying the management of Florida's Everglades for many years and this excerpt from the summary of a report by him and his colleagues expresses very well how it can play out:

> ... the current governance and management system is in a hierarchy or rigidity trap. The existing complex of institutions and actors has maintained an ongoing conflict over water use for at least forty years. This conflict has been stable and persistent, and illustrates a perversely resilient system. As such, the system seems incapable of moving beyond planning into practice unless some sort of crises (ecological, economic, political, or social) unlocks the stability of the system.[40]

The longer the conservation phase lasts the less resilient it becomes to external shocks, and when it eventually and inevitably collapses – after a fire, drought, insect outbreak in an ecosystem, a market or labour shock in a business – there is a short phase of chaos followed by one of re-organisation when new things are possible. New species that could never have become established in a mature ecosystem can now come into it: new ideas and a new structure can come into a business. The system then usually enters a new growth phase. However, if the collapse phase has been too severe, it may remain in a highly reduced, degraded state, unable to re-organise and recover. It will have been transformed.

The four phases of growth, conservation, collapse and re-organisation constitute adaptive cycles of change. The first two are the 'foreloop' of the cycle and are reasonably predictable. The 'backloop' of collapse and then re-organisation is fast and unpredictable.

The longer a system stays in the conservation phase, not only does the chance of a collapse increase but also the magnitude of the collapse; and a particular feature of social systems is a tendency to prolong the conservation phase. People in the system, especially those in charge, like it like that. But, if those in charge recognise what's happening, they can initiate an organised 'collapse' that allows a successful transformation. Du Pont, for example, began as a gunpowder factory in the American Civil War and, when the demand for gunpowder waned, they deliberately transformed into a petrochemical company (using their developed expertise), and latterly have transformed again into something like a sustainability company focused on renewable energy.

The Austrian economist Joseph Schumpeter described what he called periods of 'creative destruction' in economic systems. He defined it as a process that revolutionises the economic structure from within: it does not need disturbances from outside. It incessantly destroys the old system, incessantly creating a new one: a process that allows new things to emerge, in line with the notion of the collapse and re-organisation phases in adaptive cycles.

Think back to the collapse of the Roman Empire and the Berlin Wall and what preceded and followed them. They reflect rather closely the pattern of an adaptive cycle – in the case of the Wall its fall acting as the trigger for the collapse of communism.

Out of all the explorations of how systems change, resilience emerges as a blend of threshold dynamics in response to external shocks, and the internal dynamics of the adaptive cycle. Resilience is about probing the boundaries of change to learn how not to cross them and, where a system is in an undesirable state, learning how to cross back into the desirable one, or if that's not possible how to transform into a new system. Being able to orchestrate the changes depends on where the system is in the adaptive cycle. It's never a clear, clean, predictable process. It's complex and messy. There is never an optimal combination of the amounts of whatever makes up a system, the rates at which they are changing or the kinds and strengths of their connections. Goldilocks would have not been able to choose. Resilience calls for embracing uncertainty in building systems that will be safe when they fail, rather than trying to build fail-safe systems.

A way forward

The need for cooperative behaviour stands out in social-ecological systems that have maintained resilience – and the need becomes stronger as the size of the system increases. Cooperation, between groups, societies and nations, lies at the core of how to get off the growth trajectory and on to one that offers sustainable human wellbeing. It is 350 years since Thomas Hobbes wrote *Leviathan* – the 'life is solitary, poor, nasty, brutish and short' description of life in Britain during the English Civil War. His idea of a social contract – giving up some individual freedoms for allegiance to a national sovereignty that in turn delivered internal peace – was the start that lifted life in Britain to what it is today.

Getting England out of continual, debilitating, warring fiefdoms allowed people to cooperate, to get on with life and apply their ingenuity to developing and improving their lot. But that progress has scaled up and allegiance to the nation state has now become a big part of the problem. It is a major contributor to the global-scale problems we face: a world in which nations are the modern-day equivalent of fiefdoms. However, we do not need a global sovereign or government (you can just imagine some we could get!). We need rather a world in which nation states and global corporations cooperate in paying allegiance to a coordinated, global-scale set of institutions: a *strong* system of global governance based on a resilience perspective of how the global environmental-social system works.

In response to unbridled growth, rising inequality, loss of 'identity' and life-threatening environmental degradation, there has been a counter increase in religious fundamentalism. For many, neither option is acceptable but, though it's still a bit fuzzy, there is a third option: a resilience approach to advancing human wellbeing and ecological sustainability. It does not oppose economic growth *per se* and it favours all the advantages of a market economy, but it takes into account the secondary, especially long-term, consequences. It recognises that the state of the economy is an input to human wellbeing, rather than the other way around, and understands that top-down control aimed at achieving some optimal state or target does not, cannot, work.

The combination of resilience and cultural evolution is where hope lies. It opens up the possibility of guided self-organisation aimed at avoiding 'bad' futures. Resilience is not about choosing where to go: some particular optimal future. There is no such nirvana. As circumstances change, what's considered best today can soon become somewhere you don't want to go. We have no idea of what's coming in terms of new opportunities, hazards, technologies and human preferences, so, in the face of such uncertainty, the way forward is to learn how to keep what's currently wanted resilient, and as what's wanted changes and the world around changes, learn how to adapt to that or, if necessary, transform.

Resilience, then, is about choosing where *not* to go. It's about guiding, or shepherding, the system away from unacceptable paths and so enabling self-organisation among the acceptable pathways – some of them new or still emerging. In this sense, guided self-organisation is a deliberate but adaptive evolving process of avoiding trajectories headed for catastrophes, finding out how to break the 'lock-in' on them, and so allowing adaptation among and along pathways that maintain quality survival for humanity. In accordance with the behaviour of all complex adaptive systems, the process operates in a nested way across all the scales in society, from local communities up to the global. What's happening at one scale influences what happens at scales above and below.

A promising culture therefore – a third way that offers hope – is a shift from the overriding focus on either the norms of economic growth and me-me-me, or the religious alternative of life in the hereafter, to a culture of resilience with a liberal dash of serendipity, focused on wellbeing in society with an overriding requirement of avoiding catastrophic changes.

The resilience part is about enhancing our capacity for recovering and learning from shocks so as not to cross thresholds, and for guided self-organisation that will enable us to avoid getting stuck on some undesirable path.

The serendipity part embraces chance and novelty and combines it with sagacity in making choices about where, when and how to intervene to achieve the resilience part, how to avoid the hazards and seize the opportunities, and it becomes particularly important when the need for transformational change arises.

Embracing chance and uncertainty is all too often eliminated in today's tightly prescribed planning processes that are focused on risk avoidance, but its value has been recognised for a long time, exemplified by Louis Pasteur's well-known quote from over 100 years ago: 'Chance favours the prepared mind.'

The growth norm downplays chance by including it in a defined way, as probabilities of future conditions, and so it does not cater for real surprises. A resilience approach does that. It's about trying to develop systems that will be safe whatever the future brings (again, safe-fail rather than supposedly fail-safe). Given a goal of long-term human wellbeing, under its mantle of coping with uncertainty resilience gives priority to learning how to ride the world piggyback rather than trying to lead and dominate her.

Well, okay (I hear you mutter), that's all very well but how do we achieve it? This is where the earlier discussion about the evolution of norms and culture comes into play. Shifting normative behaviour in ways that are consistent with the necessary changes for the transition we need. Doing this in a variety of bottom-up ways through advancing norms such as valuing/fostering societal wellbeing and more practical issues such as using renewable energy. Advancing such changes collectively adds up to a change in overall normative behaviour and so to a change in culture. As this happens, it feeds up to those corporate players who are receptive to this change, giving them support, and eventually to recalcitrant corporations through a combination of social sanctions and increasing financial disadvantage. Driving a Humvee (a ridiculously huge four-wheel drive vehicle) around town was at one point considered cool but is now frowned upon and ridiculed in social media. Like smoking in restaurants, it's no longer acceptable. Norms can change. We need to develop a narrative around new norms that fit with a world of resilient high-value ecosystems and high human wellbeing.

Resilience as normative behaviour, as a culture, offers something all can embrace. Using providential chance occurrences and circumstances in a deliberate way to build and manage the resilience of the ecological and social systems that succour us offers a purpose and a means of building human wellbeing into the future. The focus on societal wellbeing, which ultimately confers greater individual wellbeing than me-me-me behaviour, steers the world away from catastrophe and because it is doable it offers hope. And having hope, in itself, confers resilience.

The challenging question to end with is: 'What kind of governance will enable resilience-based guided self-organisation to develop, and how can we transition to such a system of governance?' It will require actions of various kinds at multiple levels, but one thing is sure: a growing bottom-up pressure to change present cultural norms is a critical part of it, and a critical part of that will be the rapidly increasing power and use of social media.

Epilogue: What it's all about

Surprise is the greatest gift which life can grant us (Boris Pasternak)

In the course of my career, the questions I posed have changed and evolved as I learned from mistakes, and some discoveries. Among the things that emerged as being really important, there is something in particular about uncertainty. It is the hallmark of the way social-ecological systems function, how they are structured, and how they change. And it warns against the notion of Designer Ecosystems Ltd: an era that technology (gene-, nano-, information-) and short-term economics is ushering in. It is the uncertainty, messiness and unpredictability of ecosystems and social systems that enables them to cope with shocks and change, and is therefore an integral part of the way we need to live in and manage them.

The future will always be uncertain, subject to unexpected developments and, odd though it may seem, this is a comforting thought. We have good reason to be concerned about many aspects of our world and need to do what we can to avoid dystopia, but trying to define and achieve utopia is impossible. In his wonderful advice on how to live a good life, John Mortimer (*Where There's a Will*) has a chapter on 'Avoiding Utopia'.[41] While it's admirable to have utopia in mind, he says, there must be no serious danger of you ever reaching it.

To persist in a form that we like and that is good for us, the world needs to have an element of surprise, and the discernment and wisdom to take advantage of accidents and crises. Humans, ecosystems and human-social systems need to be disturbed and undergo periodic change if they are to absorb the pressures and disturbances that will later confront them. To foster their resilience it is necessary to probe their boundaries.

People do not like change. We seek certainty in our lives, especially as we age, but a measure of uncertainty is an essential ingredient keeping us on our toes and generating novelty and renewal. A hundred years ago, designers of utopia would not have been able to forecast another world war, nuclear bombs, the Spanish flu, the Great Depression of the 1930s or the 2008 Financial Crisis, air travel, space-based communication satellites or computers, the IT revolution and the power of social media, AIDS, bionics, 3D printing, genetic engineering or the climate change implications of an increase in atmospheric carbon dioxide. Yet the utopia they attempted to construct would have had to deal with them, and without all the apparently unnecessary variability, the response diversity, the reserve problem-solving capacity that uncertainty generates, it would have failed. Uncertainty in all parts of our linked systems of humans and nature is the essential ingredient for keeping them adaptive.

We cannot know all the things that will have happened and been discovered 100 years hence (less than one more second in the movie of Earth). But we do know that right now humanity is taking more than Earth can continue to supply and, despite a growing unease, the dominant paradigm is still short-term command-and-control towards a designed future, based on increased growth and getting more out of the world more efficiently. We also know that, to start living within our means, we need to change our goals and our ethics, our norms, from a 'me' to a 'we' society. It follows that we need a transformational change to a resilience paradigm that will enable us to absorb the shocks and low points if we are to transition to a future in which we can sustain high human wellbeing.

Three messages come out of all that has been discussed in this book, whatever scale you're concerned about – you, society, the globe:

- *Celebrate change* – resilience is about learning how to change in order to stay the same.
- *Embrace uncertainty* – try to build systems that will be safe when they fail.
- *Don't aim for some utopia* – learn how to ride the system piggyback, guiding it away from thresholds that lead into undesirable futures from which you may not be able to recover.

My study at home looks out on to a patio covered in part by a Banks' rose tree, shaped a bit like a flat-topped acacia. A bird feeder tray hangs under it, and as I sit and write I watch Eric and Freda eating sunflower seeds while looking at me through my study window. They are a pair of beautiful king parrots, regular visitors, named by my wife Laura after my parents. If there are no seeds in the tray, Eric sits on the back of a patio chair with his two-toned green wings and magnificent red chest, and shouts at me through the window, so I fetch the bag of seeds. Eric and Freda are dominant over several pairs of crimson rosellas, with their beautiful, tuneful calls: also regular visitors but they only get to feed if Eric and Freda let them. On occasion, a pair of gang-gang cockatoos arrives, with their creaky calls and his crazy red feather hair-do. There's a splendid mix of aerobatic colours and noise in a stoush with Eric and Freda that the gang-gangs generally win. Discovering what we need to do so my grandchildren's hearts can be warmed by watching Eric's and Freda's descendants interact with the rosellas and gang-gangs of their day, and helping make those needs part of the ethical and policy world in which we live, is surely part of what it's all about.

Glossary

adaptive capacity – the capacity of a system to change and re-organise in response to a disturbance or a changing environment.

adaptive cycle – the progression of complex systems, ecosystems and social systems, through a four-phase cycle. A foreloop of growth and then conservation and becoming inflexible, triggering a collapse into a backloop of chaotic unravelling followed by a phase of re-organisation, leading into a new cycle.

Anthropocene – Earth's current epoch (informally defined), starting at about the time of the Industrial Revolution, when humans began to influence the climate.

alluvium – fertile, silty soil adjacent to large rivers, deposited during floods.

billabong – a pool of water created by isolation of a bend in a river, typical in northern Australia.

bolus – a rounded lump of faeces, as expelled by elephants.

brumby (Australian English) – a wild horse.

calici virus – a virus that causes a disease in rabbits, damaging the internal organs and leading to death.

capital stocks – the stocks of natural, human, and built capital that constitute the wealth of a country.

catena – a sequence of soil types from a ridge down to the flat area at the bottom.

CSIRO – The Commonwealth Scientific and Industrial Research Organisation of Australia.

doro (Shona) – a beer made from maize, sorghum or other small grains *and* a cultivated garden at the edge of a vlei (see 'vlei' below), distinguished by the inflection on the second syllable.

food pyramid – the progressive decline in biomass up the food chain in an ecosystem, from plants through herbivores and up to the top carnivores.

GDP – gross domestic product; the monetary value of all the finished goods and services produced within a country's borders in a specific time period.

Holocene – Earth's current epoch, which began about 12 000 years ago and is characterised by an unusually stable warm climate, which allowed human civilisations to evolve.

inclusive wealth – an index based on the sum of natural, human and built capital stocks reflecting the wealth of a nation, as opposed to GDP which measures only flows of money.

impala – medium-sized antelope common in southern Africa.

inequality – economic inequality is the difference in incomes in a society, between the rich and the poor, usually measured as the 'Gini' coefficient (a statistical dispersion index of income distribution). Social inequality is the unequal opportunities and rewards in a society.

inequity – similar to inequality but with the connotation of unfairness.

keystone species – a single species on which many other species depend and which maintains an essential ecosystem function.

lobola (Shona) – bride price.

meme – an element of a culture, such as an idea or a system of behaviour passed from one individual to another, by imitation. It was coined by Richard Dawkins to highlight the difference between this kind of inheritance and the inheritance of genes.

mutational meltdown – accumulation of harmful mutations in a small population leading to loss of fitness and further decline in population size.

mudzimu bull (Shona) – a bull into which the spirit of a deceased has been placed.

myxomatosis – a viral disease that kills only rabbits.

niche – the ecological role (e.g. 'small predator') or the environmental space (e.g. 'margins of wetlands') occupied by a species in an ecosystem.

Pleistocene – the period (epoch) in Earth's history from about 2.5 million years ago until about 12 000 years ago that includes all the recent glaciations.

resilience – the ability of a system (a body, an ecosystem, a city) to absorb a disturbance and re-organise so as to keep functioning in the same kind of way – to have the same 'identity'.

response diversity – different ways of doing the same thing (performing the same function), each with different responses to kinds of disturbances.

rinderpest – a lethal viral disease of ruminants that swept through Africa around the turn of the 20th century.

sabuku (Shona) – village head.

saduna (Shona) – the head of a group of villages.

slough – a depression filled with water in the prairie provinces of Canada, created by the retreating ice cap.

social capital – the capacity for people to work together for their common interest; mostly determined by strengths of networks, leadership and trust.

spoor – the distinctive imprint of an animal's foot.

sudds – floating mats of vegetation in a swamp.

threshold (tipping point) – a critical amount of a component or rate of a process in a system that results in a change in the way the system functions and therefore in its composition.

tipping point – as for threshold, the term used mostly by social scientists.

trickle-down effect – the supposed increase in economic performance of those at low levels of economic status due to increasing the wealth of those at the top; disproved several times by prominent economists.

trophic cascade – the top-down effect of predators in an ecosystem through to the plants that alters the amounts of species in each of the layers in the food pyramid.

vlei – southern African word (Afrikaans origin) for a grass-dominated marshy area draining into a stream.

Endnotes

1   Theme song from the 1966 movie *Alfie*. Lyrics reproduced with permission from Silva Screen.

2   Note: a glacial period is not an 'ice age'. The world is now in a long-term ice age because it has ice over the poles and glaciers, but it goes through lots of glacial/interglacial phases as the extents of these wax and wane.

3   Michelle de Kretser *pers. comm.*

4   Kahneman D (2011) *Thinking, Fast and Slow*. Farrar, Straus and Giroux, New York, USA.

5   Perlin J (1989) *A Forest Journey: The Story of Wood and Civilization*. The Countryman Press, Woodstock VT, USA.

6   https://www.footprintnetwork.org/our-work/earth-overshoot-day/

7   Coulson J (2017) *9 Ways to a Resilient Child*. HarperCollins, Sydney.

8   Hirshfield J (2002) Optimism. In *Given Sugar, Given Salt*. HarperCollins, New York.

9   Goldsmith O (1990) *Oliver Goldsmith's History of the Natural World*. Studio Editions, London, UK.

10 Holling CS (1966) The functional response of invertebrate predators to prey density. *Memoirs of the Entomological Society of Canada* **48**, 1–86.

11 Holling CS (1973) Resilience and stability of ecological systems. *Annual Review of Ecology and Systematics* **4**, 1–23.

12 Smita Malhotra, MD. August 2014, *Huffington Post* blog.

13 Wolin SJ, Wolin S (1993) *The Resilient Self: How Survivors of Troubled Families Rise Above Adversity*. Villard Books, New York, USA.

14 Fletcher D, Sarkar M (2013) Psychological resilience: a review and critique of definitions, concepts and theory. *European Psychologist* **18**, 12–23.

15 Camus A (1954) Return to Tipasa. In *Myth of Sisyphus and Other Essays*. Reissue edn 1991. pp. 201–202. Vintage, New York, USA.

16 Sturgeon JA, Zautra AJ (2010) Resilience: a new paradigm for adaptation to chronic pain. *Current Pain and Headache Reports* **14**(2), 105–112.

17 Ostrom E (1990) *Governing the Commons: The Evolution of Institutions for Collective Action*. Cambridge University Press, Cambridge, UK.

18 Hardin G (1968) The tragedy of the commons. *Science* **162**, 1243–1248.

19 Walker B, Salt D (2010) *Resilience Practice*. Island Press, Washington DC, USA.

20 Hahn T, Olsson P, Folke C, Johannson K (2006) Trust-building, knowledge generation and organizational innovations: the role of a bridging organization for adaptive comanagement of a wetland landscape around Kristianstad, Sweden. *Human Ecology* **34**(4), 573–592.

21 Gappah P (2015) On translating Orwell's Animal Farm. In *English PEN*. <https://www.englishpen.org/pen-atlas/on-translating-orwells-animal-farm/>

22 Grayling AC (2017) *Democracy and its Crisis*. Oneworld Publications, London UK.

23 Buldyrev SV, Parshani R, Paul G, Stanley HE, Havlin S (2010) Catastrophic cascade of failures in interdependent networks. *Nature* **464**, 1025–1028.

24 Keck M, Sakdapolrak P (2013) What is social resilience? Lessons learned and ways forward. *Erkunde* **67**, 5–19.

25 Tainter J (1990) *Collapse of Complex Societies*. New Studies in Archaeology. Cambridge University Press, Cambridge, UK.

26 'What's it all about, Alfie' was sung by Cher in the 1966 original and made famous in Europe by Cilla Black.

27 Meadows DH, Meadows DL, Randers J, Behrens III WW (1972) *The Limits to Growth: A Report for the Club of Rome's Project on the Predicament of Mankind*. Universe Books, New York, USA.

28 Quiggin J (2010) *Zombie Economics: How Dead Ideas Still Walk Among Us*. Princeton University Press, Princeton NJ, USA.

29 UNU-IHDP, UNEP (2014) *Inclusive Wealth Report 2014: Measuring Progress Toward Sustainability*. Cambridge University Press, Cambridge, UK.

30 Mill JS (1848) *Principles of Political Economy*. John W. Parker, London, UK.

31 Wilkinson R, Pickett K (2009) *The Spirit Level: Why More Equal Societies Almost Always Do Better*. Allen Lane, London, UK.

32 Putnam RD (2000) *Bowling Alone: The Collapse and Revival of American Community*. Simon & Schuster, New York, USA.

33 Sukhdev P (2012) *Corporation 2020: Transforming Business for Tomorrow's World*. Island Press, Washington DC, USA.

34 Costanza R, Caniglia E, Fioramonti L, Kubiszewski I, Lewis H *et al.* (2018) Towards a sustainable wellbeing economy. *Solutions* **9**(2). <https://www.thesolutionsjournal.com/article/toward-sustainable-wellbeing-economy/>

35 Leopold A (1949) *A Sand County Almanac, and Sketches Here and There*. Oxford University Press, New York, USA.

36 Bichieri C (2006) *The Grammar of Society: The Nature and Dynamics of Social Norms*. Cambridge University Press, Cambridge, UK.

37 Walker BH, Salt D (2006) *Resilience Thinking: Sustaining Ecosystems and People in a Changing World*. Island Press, Washington DC, USA.

38 Levin S (1999) *Fragile Dominion: Complexity and the Commons*. Perseus Books, New York, USA.

39 Brown P (2015) *Missing Up*. Vagabond Press, Sydney.

40 Gunderson L, Cosens BA, Chaffin BC, Tom Arnold CA, Fremier AK, Garmestani AS, *et al.* (2017) Regime shifts and panarchies in regional scale social-ecological water systems. *Ecology and Society* **22**(1), 1–31.

41 Mortimer J (2003) *Where There's a Will*. Viking, London, UK.

Index